Book Series on Theory and Technology of Intelligent Manufacturing and Robotics

Editors-in-Chief: Han Ding & Ronglei Sun

Kun Bai
Kok-Meng Lee

Permanent Magnet Spherical Motors

Model and Field Based Approaches for Design, Sensing and Control

图书在版编目(CIP)数据

永磁球形电机:基于模型以及物理场的设计、传感和控制＝Permanent Magnet Spherical Motors:Model and Field Based Approaches for Design,Sensing and Control:英文/白坤,李国民著.—武汉:华中科技大学出版社,2019.1
(智能制造与机器人理论及技术研究丛书)
ISBN 978-7-5680-3902-4

Ⅰ.①永… Ⅱ.①白… ②李… Ⅲ.①球形-永磁式电机-英文 Ⅳ.①TM351

中国版本图书馆CIP数据核字(2018)第233935号

Sales in Mainland China Only
本书仅限在中国大陆地区发行销售

永磁球形电机:基于模型以及物理场的设计、传感和控制
YONGCI QIUXING DIANJI:JIYU MOXING YIJI WULICHANG DE
SHEJI CHUANGAN HE KONGZHI

白 坤 李国民 著

丛书策划：王连弟 俞道凯
策划编辑：俞道凯
责任监印：周治超

出版发行：华中科技大学出版社(中国·武汉)   电话：(027)81321913
　　　　　武汉市东湖新技术开发区华工科技园   邮编：430223
录　　排：武汉三月禾文化传播有限公司
印　　刷：湖北新华印务有限公司
开　　本：710mm×1000mm 1/16
印　　张：11.5
字　　数：239千字
版　　次：2019年1月第1版第1次印刷
定　　价：128.00元

本书若有印装质量问题,请向出版社营销中心调换
全国免费服务热线：400-6679-118 竭诚为您服务
版权所有 侵权必究

Huazhong University of Science and Technology Press

Website: http://www.hustp.com/
Book Title: Permanent Magnet Spherical Motors Model and Field Based Approaches for Design, Sensing and Control

Copyright @ 2019 by Huazhong University of Science & Technology Press. All rights reserved. No part of this publication may be reproduced, stored in a database or retrieval system, or transmitted in any form or by any electronic, mechanical, photocopy, or other recording means, without the prior written permission of the publisher.

Contact address: No. 6 Huagongyuan Rd, Huagong Tech Park, Donghu High-tech Development Zone, Wuhan City 430223, Hubei Province, P.R. China.
Phone/fax: 8627-81339688; **E-mail:** service@hustp.com

Disclaimer

This book is for educational and reference purposes only. The authors, editors, publishers and any other parties involved in the publication of this work do not guarantee that the information contained herein is in any respect accurate or complete. It is the responsibility of the readers to understand and adhere to local laws and regulations concerning the practice of these techniques and methods. The authors, editors and publishers disclaim all responsibility for any liability, loss, injury, or damage incurred as a consequence, directly or indirectly, of the use and application of any of the contents of this book.

First published: 2019
ISBN: 978-7-5680-3902-4

Cataloguing in publication data:
A catalogue record for this book is available from the CIP-Database China.

Printed in The People's Republic of China

Book Series on Theory and Technology of Intelligent Manufacturing and Robotics

Consultative Group of Experts

Chairman Youlun Xiong, Huazhong University of Science and Technology, Wuhan, China

Members

Bingheng Lu, Xi'an Jiaotong University, Xi'an, China
Xueyu Ruan, Shanghai Jiao Tong University, Shanghai, China
Jianwei Zhang, Universität Hamburg, Hamburg, Germany
Jianying Zhu, Nanjing University of Aeronautics and Astronautics, Nanjing, China
Zhuangde Jiang, Xi'an Jiaotong University, Xi'an, China
Di Zhu, Nanjing University of Aeronautics and Astronautics, Nanjing, China
Huayong Yang, Zhejiang University, Hangzhou, China
Xinyu Shao, Huazhong University of Science and Technology, Wuhan, China
Zhongqin Lin, Shanghai Jiao Tong University, Shanghai, China
Jianrong Tan, Zhejiang University, Hangzhou, China

Advisory Panel

Chairman Kok-Meng Lee, Georgia Institute of Technology, Atlanta, GA, USA

Members

Haibin Yu, Shenyang Institute of Automation Chinese Academy of Sciences, Shengyang, China
Tianmiao Wang, Beihang University, Beijing, China
Zhongxue Gan, Fudan University, Shanghai, China
Xiangyang Zhu, Shanghai Jiao Tong University, Shanghai, China
Lining Sun, Soochow University, Suzhou, China
Guilin Yang, Ningbo Institute of Industrial Technology, Ningbo, China
Guang Meng, Shanghai Academy of Spaceflight Technology, Shanghai, China
Tian Huang, Tianjin University, Tianjin, China
Feiyue Wang, Institute of Automation, Chinese Academy of Sciences, Beijing, China
Zhouping Yin, Huazhong University of Science and Technology, Wuhan, China
Tielin Shi, Huazhong University of Science and Technology, Wuhan, China
Hong Liu, Harbin Institute of Technology, Harbin, China
Bin Li, Huazhong University of Science and Technology, Wuhan, China
Dan Zhang, Beijing Jiaotong University, Beijing, China
Zhongping Jiang, New York University, NU, USA
Minghui Huang, Central South University, Changsha, China

Editorial Committee

Chairman Han Ding, Huazhong University of Science and Technology, Wuhan, China
Ronglei Sun, Huazhong University of Science and Technology, Wuhan, China

Members

Cheng'en Wang, Shanghai Jiao Tong University, Shanghai, China
Yusheng Shi, Huazhong University of Science and Technology, Wuhan, China
Shudong Sun, Northwestern Polytechnical University, Xi'an, China
Dinghua Zhang, Northwestern Polytechnical University, Xi'an, China
Dapeng Fan, National University of Defense Technology, Changsha, China
Bo Tao, Huazhong University of Science and Technology, Wuhan, China
Yongcheng Lin, Central South University, Changsha, China
Zhenhua Xiong, Shanghai Jiao Tong University, Shanghai, China
Yongchun Fang, Nankai University, Tianjin, China
Hong Qiao, Institute of Automation, Chinese Academy of Sciences, Beijing, China
Zhijiang Du, Harbin Institute of Technology, Harbin, China
Xianming Zhang, South China University of Technology, Guangzhou, China
Xinjian Gu, Zhejiang University, Hangzhou, China
Jianda Han, Nankai University, Tianjin, China
Gang Xiong, Institute of Automation, Chinese Academy of Sciences, Beijing, China

Editor

▶ **Kun Bai** received his B.S. degree from Zhejiang University, China in 2006 and earned his M. S. and Ph. D. degrees from the Woodruff School of Mechanical Engineering at Georgia Institute of Technology, Atlanta, US in 2009 and 2012 respectively. Currently, he is Associate Professor with the State Key Laboratory of Digital Manufacturing Equipment and Technology and the School of Mechanical Science and Engineering at Huazhong University of Science and Technology, China.

Dr. Bai's research areas include actuators, sensing and control systems, in which he has published over 30 peer-viewed papers and held over 10 patents. He has extensive expertise and experience in developing spherical motors for a variety of applications, ranging from manufacturing to robotics. These successful attempts have led to systems/equipments with superior advantages in terms of structural simplicity and dynamic response, comparing to their counterparts built with conventional single-axis motors. Dr. Bai has also made major contributions in developing novel multi-DOF sensing/control methods and systems which significantly improve the feedback control performance and efficiency of spherical motors.

▶ **Professor Kok-Meng** Lee earned his B.S. degree from the University at Buffalo, the State University of New York, Buffalo, NY, USA, in 1980, and M.S. and Ph.D. degrees from Massachusetts Institute of Technology, Cambridge, MA, USA, in 1982 and 1985, respectively. He is currently Professor of Mechanical Engineering at Georgia Institute of Technology, Atlanta, GA, USA. He is also Distinguished Professor with the State Key Laboratory of Digital Manufacturing Equipment and Technology, Huazhong University of Science and Technology, China, under Thousand Talents Plan.

Prof. Lee's research interests include system dynamics/control, robotics, automation, and mechatronics. He is a world renowned researcher with more than 30 years of research experience in magnetic field modeling and design, optimization and implementation of electromagnetic actuators. He has published over 150 peer-reviewed papers and holds 8 patents in machine vision, three degrees of freedom (DOF) spherical motor/encoder, and live-bird handling system. He is IEEE/ASME Fellow and was the Editor-in-Chief for the IEEE/ASME Transactions on Mechatronics from 2008 to 2013. Recognitions of his research contributions include the National Science Foundation (NSF) Presidential Young Investigator, Sigma Xi Junior Faculty Research, International Hall of Fame New Technology, and Kayamori Best Paper awards.

Preface

In recent years, "Intelligent manufacturing+tri-co robots" are particularly eye-catching, presenting the characteristics of the era of the perception of things,the interconnect of things,the intelligence of thing.Intelligent manufacturing and tri-co robots industry will be the strategic emerging industry with priority development,it is also a huge engine for "Made in China 2049". It's remarkable that the large-scale tri-co robots industry formed by smart cars ,drones and underwater robots will be a strategic areas of countries to compete in the next 30 year, and have influence on economic development, social progress, and war forms. The related manufacturing sciences and robotics are comprehensive disciplines that links and covers material sciences, information sciences, and life sciences. Like other engineering sciences and technical sciences, tri-co robots industry also will be the big sciences that provide a way to understand and transform the world. In the mid-20th century,the publication of Cybernetics and Engineering Cybernetics created a new era of engineering sciences.Since the 21th century,the manufacturing sciences, robotics and artificial intelligence and other fields have been extremely active and far-reaching,they are the sources of the innovation of "Intelligent manufacturing+ tri-co robots".

Huazhong University of Science and Technology Press follows the trend of the times, aiming at the technological frontiers of intelligent manufacturing and robots, organizes and plans this series of Intelligent Manufacturing and Robot Theory & Technology Research Series .The series covers a wide range of topics,experts and professors are warmly welcome to write books from different perspectives,different aspects, and different fields.The key points of the topics include but are not limited to:the links of intelligent manufacturing,such as research, development, design, processing, molding and assembly, etc;the fields of intelligent manufacturing,such as intelligent control, intelligent sensing, intelligent equipment, intelligent systems, intelligent logistics and intelligent automation, etc;development and application of robots,such as industrial robots, service robots, extreme robots, land-sea-air robots, bionics/artificial/robots, soft robots and micro-nano robots;artificial intelligence, cognitive science, big data, cloud manufacturing, Internet of things and Internet,etc.

This series of books will become a platform for academic exchange and cooperation between experts and scholars in related fields, a zone where young scientists thrive, and an international arena for scientists to display their research results.Huazhong University of Science and Technology Press will cooperate with international academic publishing organizations such as Springer Publishing House to publish and distribute this series of books. Also, the company has established close with relevant international academic conferences and journals, creating a good environment to enhance the academic level and practical value,expand the international influence of the series.

In recent years, people from all walks of life, university teachers and students, experts, scientists and technicians in various fields are more and more enthusiastic about intelligent manufacturing and robotics.This series of books will become the link between experts, scholars, university teachers and students and technicians, enhance the connection between authors, editors and readers, speed up the process of discovering, imparting , increasing and updating knowledge, contribute to economic construction, social progress, and scientific and technological development.

Finally,I sincerely thank the authors, editors and readers who have contributed to this series of books,for adding,gathering,and exerting positive energy for innovation-driven development, thank the relevant personnel of Huazhong University of Science and Technology Press for their hard work in the process of organizing and scheming of the series of book.

Professor of Huazhong University of Science and Technology
Academician of Chinese Academy of Sciences

Youlun Xiong
September, 2017

Preface

Rapid advances of intelligent machines for smart manufacturing equipment, driverless vehicles, robotics, and medical industries continue to motivate new designs and applications of multi-degree-of-freedom (DOF) actuators capable of complex motion and precise force/torque manipulations to complete tasks that have never been automated before. Extensive efforts to develop novel actuators with compact designs and dexterous manipulations can be found in both academic research and industrial development. Unlike multi-DOF systems with designs based on bulky serial/parallel combinations of single-axis spin motors and transmission mechanisms, spherical motors/actuators are direct-drive and can achieve multi-DOF rotational motions in a single ball joint. Because of these attractive features, along with the structural simplicity and the capability to achieve quick singularity-free motion, spherical motors are expected to play a significant role in the development of intelligent machines.

In this book, we provide fundamentals for practical designs of spherical motors with the intent to push forward the development of high-performance spherical motors. This book is organized into three parts: The first part begins with the methods for modeling the three-dimensional (3D) electromagnetic fields involved in a spherical motor, and the multi-dimensional forces and torques generated electromechanically between its rotor and stator. The second part presents the sensing techniques for measuring the multi-DOF joint motion in real time. The third part offers methods for controlling the coupled rotational motions of spherical motors. While this book is primarily intended for students, researchers, and engineers studying/developing spherical motors, those who work in the area of electric machines should find the modeling, sensing and control methods presented here relevant to the development of various electromagnetic motion systems.

This book is an outcome of the research work accomplished by the authors on spherical motors over the years in Georgia Institute of Technology (USA) and Huazhong University of Science and Technology (China). The authors sincerely appreciate the institutional support received from the two organizations. Part of the work on DMP model and orientation sensing has also built on research works by many former students, particularly Dr. Hungsun Son and Dr. Shaohui Foong, during their studies at Georgia Institute of Technology.

Wuhan, China
Atlanta, USA

Kun Bai
Kok-Meng Lee

Nomenclature

Symbols (Uppercase)

A	Magnetic vector potential
B	Magnetic flux density
H	Magnetic field intensity
I	Moment of inertia
J	Current density
[K]	Torque characteristic matrix
K	Torque characteristic vector
$\mathbf{K_P}, \mathbf{K_I}, \mathbf{K_D}$	Proportional, integral, and derivative gain matrices
M	Magnetization
M_0	Magnetization strength
[M]	Inertia matrix
N_E	Number of electromagnets
N_P	Number of permanent magnets
N_W	Number of turns in the winding of an EM
S	Sensor index vector
T	Motor torque
XYZ	Stator coordinate frame
Γ	Bijective domain
Ω	MFD-defined domain
Φ	Magnetic scalar potential
Λ	Magnetic flux linkage

Symbols (Lowercase)

g	Gravity torque vector
i	EM/coil current input
m	Magnetic dipole strength
n	Normal vector
q	Rotor orientation

u	Current input vector
xyz	Rotor coordinate
α, β, γ	XYZ Euler angles
θ, ϕ, r	Spherical coordinates in rotor frame
λ	Pole separation angle
μ	Permeability of magnetic material
μ_0	Permeability of free space (air)

Abbreviations

DC	Direct current
DFC	Direct field-feedback control
DMP	Distributed multi-pole
DOF	Degree of freedom
EM	Electromagnetic magnet
MFD	Magnetic flux density
PM	Permanent magnet
PMSM	Permanent magnet spherical motor
TCV	Torque characteristic vector
WCR	Weight-compensating regulator

Contents

1 Introduction 1
   1.1 Background 1
   1.2 The State of the Art 3
       1.2.1 Magnetic Modeling and Analysis 6
       1.2.2 Orientation Sensing 8
       1.2.3 Control Methods 10
   1.3 Book Outline 12
   References 14

Part I Modeling Methods

2 General Formulation of PMSMs 21
   2.1 PMSM Electromagnetic System Modeling 21
       2.1.1 Governing Equations of Electromagnetic Field 21
       2.1.2 Boundary Conditions 24
       2.1.3 Magnetic Flux Linkage and Energy 25
       2.1.4 Magnetic Force/Torque 26
   2.2 PMSM Rotor Dynamics 27
   References 30

3 Distributed Multi-pole Models 31
   3.1 Distributed Multi-pole Model for PMs 31
       3.1.1 PM Field with DMP Model 32
       3.1.2 Numerical Illustrative Examples 35
   3.2 Distributed Multi-pole Model for EMs 43
       3.2.1 Equivalent Magnetization of the ePM 45
       3.2.2 Illustrations of Magnetic Field Computation 47
   3.3 Dipole Force/Torque Model 47
       3.3.1 Force and Torque on a Magnetic Dipole 47
       3.3.2 Illustration of Magnetic Force Computation 49

	3.4	Image Method with DMP Models	52
		3.4.1 Image Method with Spherical Grounded Boundary	53
		3.4.2 Illustrative Examples	56
		3.4.3 Effects of Iron Boundary on the Torque	58
	3.5	Illustrative Numerical Simulations for PMSM Design	62
		3.5.1 Pole Pair Design	65
		3.5.2 Static Loading Investigation	70
		3.5.3 Weight-Compensating Regulator.....................	71
	References		79
4	**PMSM Force/Torque Model for Real-Time Control**		81
	4.1	Force/Torque Formulation.................................	81
		4.1.1 Magnetic Force/Torque Based on the Kernel Functions ...	82
		4.1.2 Simplified Model: Axis-Symmetric EMs/PMs	85
		4.1.3 Inverse Torque Model	86
	4.2	Numerical Illustrations	86
		4.2.1 Axis-Asymmetric EM/PMs.........................	86
		4.2.2 Axis-Symmetric EM/PM	90
	4.3	Illustrative PMSM Torque Modelling.......................	93

Part II Sensing Methods

5	**Field-Based Orientation Sensing**		99
	5.1	Coordinate Systems and Sensor Placement..................	99
	5.2	Field Mapping and Segmentation	100
	5.3	Artificial Neural Network Inverse Map	102
	5.4	Experimental Investigation	103
	References		107
6	**A Back-EMF Method for Multi-DOF Motion Detection**		109
	6.1	Back-EMF for Multi-DOF Motion Sensing	109
		6.1.1 EMF Model in a Single EM-PM Pair	111
		6.1.2 Back-EMF with Multiple EM-PM Pairs	112
	6.2	Implementation of Back-EMF Method on a PMSM	114
		6.2.1 Mechanical and Magnetic Structure of the PMSM	115
		6.2.2 Numerical Solutions for the MFL Model...............	116
		6.2.3 Experiment and Discussion	118
		6.2.4 Parameter Estimation of the PMSM with Back-EMF Method..	120
	References		122

Part III Control Methods

7 Direct Field-Feedback Control 125
   7.1 Traditional Orientation Control Method for Spherical Motors 125
        7.1.1 PD Control Law and Stability Analysis 126
        7.1.2 Comments on Implementation of Traditional Control Methods 127
   7.2 Direct Field-Feedback Control 128
        7.2.1 Determination of Bijective Domain 129
        7.2.2 DFC Control Law and Control Parameter Determination 129
        7.2.3 DFC with Multi-sensors 130
   7.3 Numerical 1-DOF Illustrative Example 131
        7.3.1 Sensor Design and Bijective Domain Identification 131
        7.3.2 Field-Based Control Law 133
        7.3.3 Numerical Illustrations of Multiple Bijective Domains 135
   7.4 Experimental Investigation of DFC for 3-DOF PMSM 135
        7.4.1 System Description 135
        7.4.2 Sensor Design and Bijective Domains 138
        7.4.3 Bijective Domain 139
        7.4.4 TCV Computation Using Artificial Neural Network (ANN) 142
        7.4.5 Experimental Investigation 142
   References 150

8 A Two-Mode PMSM for Haptic Applications 151
   8.1 Description of the PMSM Haptic Device 151
        8.1.1 Two-Mode Configuration Design for 6-DOF Manipulation 153
        8.1.2 Numerical Model for Magnetic Field/Torque Computation 154
        8.1.3 Field-Based TCV Estimation 155
   8.2 Snap-Fit Simulation 156
        8.2.1 Snap-Fit Performance Analyses 158
        8.2.2 Snap-Fit Haptic Application 159
   References 164

Chapter 1
Introduction

Spherical motors which can achieve multi-DOF rotational motion in one joint have wide potential applications in modern manufacturing, robotics, automobile and medical industries. To provide systematic approaches for developing high-performance spherical motors, the basic concepts and solutions for the modeling, sensing and control are the focus of this book. This chapter begins with the motivations along with the exemplary applications of spherical motors, which is followed by a state-of-the-art review and a brief introduction of some fundamental concepts and operational principles of spherical motors. Finally, the outline of the book is presented.

1.1 Background

Multi-degree-of-freedom (DOF) actuators are widely used in industry, particularly in emerging applications where the end-effectors must be re-oriented smoothly, rapidly and precisely. In modern manufacturing industries, the trend to downscale equipment for manufacturing products on "desktops" has motivated the development of compact mechatronic platforms capable of performing various machining tasks. Figure 1.1a shows a micro-factory system [1, 2] which consists of a high-speed spindle cutter and a multi-DOF rotational stage. The multi-DOF actuator provides dexterous motions to manipulate the work-piece mounted on the stage for sophisticated machining by the cutter. Similar manipulation can also be found in conformal printing on 3D/flexible surfaces (Fig. 1.1b) where the substrate must be continuously re-orientated to align the surface normal with the print head direction, which allows conformal printing of circuits onto 3D substrates with a high level of surface topography [3].

Besides manufacturing applications, multi-DOF actuators are essential components in medical fields. Figure 1.1c shows a handheld micromanipulator for surgery. The end-effector (needle) must be controlled in up to 6-DOF (both position

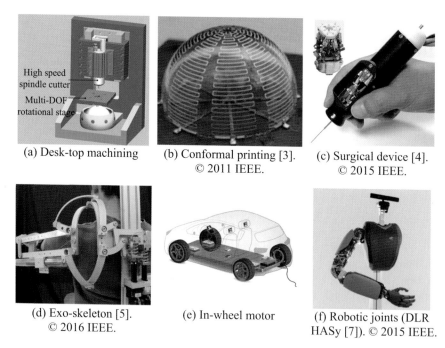

(a) Desk-top machining (b) Conformal printing [3]. © 2011 IEEE. (c) Surgical device [4]. © 2015 IEEE.

(d) Exo-skeleton [5]. © 2016 IEEE. (e) In-wheel motor (f) Robotic joints (DLR HASy [7]). © 2015 IEEE.

Fig. 1.1 Multi-DOF systems

and orientation) in order to actively compensate the hand tremor during surgical operations [4]. In Fig.1.1d, an exo-skeleton which accommodates the joint motion of human shoulders can be used in the rehabilitation process for post-stroke patients [5]. The actuator systems must offer dexterous motion and direct force/torque manipulation in order to actively cooperate with the human motion. Apart from these applications, multi-DOF actuators with novel topology and advanced control efficiencies continue to play a leading role in advancing the development of many modern systems such as automobile in-wheel motors [6] in Fig. 1.1e and robotic joint motors [7] in Fig. 1.1f.

Although multi-DOF motions can be generated by means of multiple single-axis spin-motors connected in serial or in parallel with external linkages or transmission mechanisms, such systems have some intrinsic disadvantages:

- The structural complexity of the multi-DOF mechanism, which often results in kinematic singularities in their workspace, greatly degrades its motion dexterity.
- The added mass and moment-of-inertia associated with the moving linkages for connecting the multiple single-axis actuators in a multi-DOF mechanism are the primary causes of its bulky size and poor dynamic performance.
- The friction between the moving parts results in wears which, along with the backlash in the motion transmission mechanisms (such as gears, timing belts

1.1 Background

and external linkages between the motors and the end-effectors), seriously affect motion control accuracies and robustness in practical applications.
- Furthermore, the (friction and backlash) nonlinearities in the transmission mechanisms make the force/torque manipulation extremely difficult to control, which are essential for applications like rehabilitation and haptic uses.

To overcome these problems, a number of novel actuators have been developed. Among these are the ball-joint-like spherical motors capable of providing 3-DOF rotational motions in a single joint. By eliminating the transmission mechanisms (and thus the associated frictions and moment-of-inertia), the spherical motor is simple/compact in structure, has no singularities except at the boundaries of its workspace and thus can offer continuous rotational motion essential for dexterous manipulation tasks with high dynamic performance. As the driving forces and torques are directly applied on the rotor, the spherical motors can be used for applications where precise force/torque manipulation of the end-effector is required. Originally proposed in [8] as a robotic wrist actuator, a variety of spherical motors have been developed since then. Recent spherical-motor applications include actuation for digital cameras [9], robotic joints [10], machining stages [11], satellite attitude control [12], a haptic interface for snap-fit designs [13], and a traction motor for an electrical wheelchair [14].

Permanent magnet spherical motors (PMSMs) take many forms and have been studied by many researchers in the past decades. Inheriting the merits of spherical motors capable of offering multi-DOF rotational motions in a single joint, PMSMs incorporating PM poles in the rotor have the advantages such as high force/torque density (due to the strong magnetic field of rare-earth PMs), brushless and wire-free rotor design and compact size. PMSMs have drawn more and more attention, and are expected to be employed in a variety of industrial applications. In this book, we focus on the fundamental studies and technical issues from the perspectives of modeling, sensing and control of a PMSM. The materials offered here are essential bases for developing practical PMSMs; their effective applications will contribute to and benefit manufacturing, robotics, automobile as well as medical industries.

1.2 The State of the Art

Spherical motors take a number of forms which can be categorized into electromagnetic, piezoelectric (or ultrasonic) [15, 16] and cable/wheel-driven systems [17, 18]. Most of the spherical motors are based on the principle of electromagnetism, which include induction [19–25], direct current (DC) [26, 27], stepper [28–30], variable-reluctance (VR) [31, 32], and permanent magnet spherical motors (PMSMs) [33, 34]. The earliest form of electromagnetic spherical motors was designed as an induction motor. The first induction motor in spherical form with a tiltable angle between stator and rotor axes was introduced by Williams et al. [19]. The spin shaft of the motor was fixed (thus with 1-DOF rotor motion) and the stator blocks can be pre-tilted in order to adjust the spin speed range. The spherical motor concept with multi-DOF motion was first proposed by Vachtsevanos et al. [20] for robotic manipulators. The wrist-like motor was composed of a ball rotor and a socket-like stator and was designed to achieve 3-DOF of motion in a single joint for purposes of dexterous actuation. The magnetic fields and torques of this motor were analyzed in [21]. Foggia et al. later developed a motor capable of rotating around three independent axes [22]. More recently, a shell-like spherical induction motor [14] has been reported by Fernandes and Branco which consists of copper windings in the stator slots and a rotor composed by electrically conductive material and soft composite material to close the magnetic circuit. There are also spherical motors based on the principle of DC drive including studies by Hollis et al. [26] and Kaneko et al. [27].

The basic concept of a spherical stepper, featured with a relatively simple and compact design, was originally proposed by Lee et al. in [28, 29]. The spherical stepper with NdFe-based PMs in rotor and coils in stator offered a relatively large range of motion $\pm 45°$ and possessed isotropic properties in motion. Chirikjian and Stein developed a commutation algorithm for a spherical stepper motor [30].

With the wide availability of high-coercive rare-earth PMs at low cost, PMSMs with the advantages of high force/torque density and low mechanical wear (brushless) have been a popular research topic worldwide. Lee, Roth and Zhou extended the design concept of a spherical stepper to a variable reluctance spherical motor (VRSM, Fig. 1.2) such that high-resolution motion can be achieved with a relatively small number of rotor and stator poles [31, 32].

Lee and Son have also proposed a spherical wheel motor (SWM, Fig. 1.3) [33, 34] with decoupled rotor motions into 2-DOF inclination and 1-DOF shaft spinning; and the control performances have been tested.

Yan et al. developed an experimental prototype to verify the analytical torque model of a PM spherical actuator [35–37] consisting of tapered stator coils and dihedral-shaped rotor PMs (Fig. 1.4). Based on the analytical model that relates the motor torque and the design parameters, the parametric effects (such as geometries

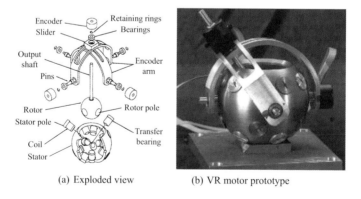

Fig. 1.2 Variable reluctance spherical motor by Lee et al. [32]

1.2 The State of the Art

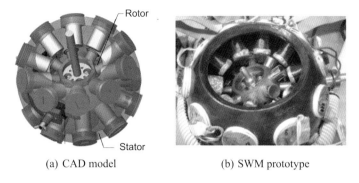

(a) CAD model (b) SWM prototype

Fig. 1.3 Spherical wheel motor by Lee et al. [34]

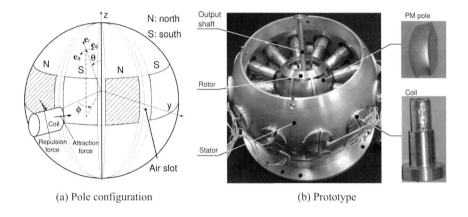

(a) Pole configuration (b) Prototype

Fig. 1.4 PM spherical actuator by Yan et al. [36]

of coils and PM poles, and structure materials) on the torque output were investigated, which led to a PMSM design with stacked cylindrical PMs as the rotor poles [38] to facilitate the fabrication.

In [39], Kim et al. have proposed the design and implementation methods for a spherical actuator that can generate two tilt-motion degrees of freedom (Fig. 1.5). Voice coil motors were adopted as actuators in this spherical actuator to utilize their simple driving principle and constant torque coefficient characteristic. The sensing and guiding mechanism was built inside the proposed spherical actuator and enabled compactness and ease of connection to other application systems. The actuator was designed using a design optimization framework to obtain high torque and a prototype actuator was manufactured with optimally designed parameters, based on which the performance was evaluated.

Rossini et al. have presented a reaction spherical actuator (Fig. 1.6) for satellite attitude control as an innovative momentum exchange device [12, 40, 41]. The

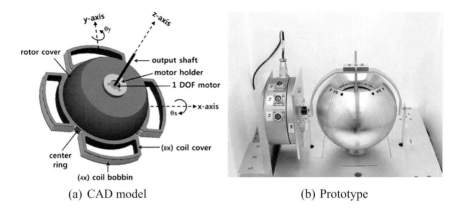

Fig. 1.5 Two-DOF spherical actuator by Kim et al. [39]. © 2015 IEEE

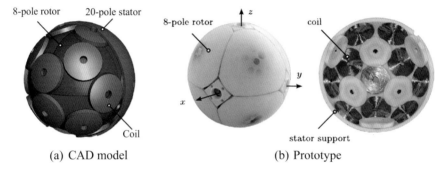

Fig. 1.6 Spherical reaction motor by Rossini et al. [40]. © 2015 Elsevier

spherical actuator is composed of an 8-pole PM spherical rotor and a 20-pole stator with electromagnets. The reaction rotor is supported by a magnetic bearing and can be electronically accelerated in any direction making all the three axes of stabilized spacecraft controllable by a unique device.

There are also a variety of works that have been conducted on the development of PMSMs. These works are introduced in the following section from the perspectives of modeling, sensing and control respectively.

1.2.1 Magnetic Modeling and Analysis

The design and control of PMSMs require the modeling and analysis of magnetic fields and forces/torques. As an electromagnetic system, the magnetic fields in a PMSM can be solved from appropriately formulated Laplace/Poisson equations,

1.2 The State of the Art

from which the magnetic force/torque between the EM and the PM poles can be calculated using with the Lorentz force law. Depending on the architecture of the designs, the force and torque models between the stator EMs and the rotor PMs can be obtained using a number of different approaches. Finite element simulations to compute the force characteristics between a stator coil and the rotor of a spherical machine with variable pole pitches were proposed in [42]. In [35], the magnetic fields of a spherical actuator was obtained by solving the Laplace equation; and the magnetic torque was derived by employing Lorentz integration. A similar approach was proposed in [12], where the analytical models for force and torque were derived by solving the Laplace equation in spherical harmonic functions, based on which closed-form linear expressions of forces and torques for all possible orientations of the rotor were obtained.

As the complexities of the solutions to Laplace equation are highly dependent on the geometries and materials in PMSM designs, and that the Lorentz force requires volume integration, the necessities to develop computation-efficient models have also motivated the development of alternative approaches to characterize magnetic fields and torque models for PMSMs. This effort has led to the distributed multi-pole (DMP) method [43] that computes the 3D magnetic field of a PM in closed form. Using the DMP method, the effects of PM geometrical parameters on the torque performance of a spherical wheel motor [33] were investigated. This work has been extended in [44] where an improved method to derive an equivalent permanent magnet (ePM) such that the magnetic field of the original multilayer EM can be characterized by a distributed set of multi-poles (DMP) model. Furthermore, with DMP models characterizing the magnetic fields of EMs and PMs (with accountability of their shapes), the dipole force equation (which was also proposed in [44]) leads to closed-form force/torque expressions by replacing integrations in commonly used Lorentz force equation or Maxwell stress tensor with algebraic summations, which dramatically reduces the computational time in design and analysis processes for PMSMs. The DMP models and dipole force equations have also been extended in [45] to take into account of ferromagnetic boundaries in PMSM designs. A generalized source modeling method, distributed multilevel current (DMC) models which utilized equivalent magnetizing currents as local point sources to describe material effects of commonly used magnetic components (such as EMs, PMs and ferromagnetic materials) was presented [46]. This method has not only provided closed-form solutions to the magnetic field and force problems but also led to an efficient means for topology optimization for electromagnetic actuators [47].

There are also lumped-parameter-based methods for formulating the PMSM magnetic field and force/torque. Lee and Kwan [29] developed torque model of a variable reluctance spherical motor and the permanence-based torque model was further investigated theoretically in [31] using finite element methods, and experimentally investigated in [32]. Even though the magnetic torque in a spherical motor is dependent on both magnetic fields and orientation, the torque models in the studies mentioned above were all formulated in terms of orientation due to the complexity and implicit relationship between magnetic fields and torque. In [48], a closed-form torque model was formulated that used curve-fitting functions to estimate torques based on the relative positions between PMs and EMs. As there are usually more current inputs than the mechanical DOFs in PMSMs, an inverse torque model to find the optimal current input minimizing the total input energy from desired torque was presented in [32]. Similar methods were also used in [49] and [50]. Other than the analysis of magnetic fields and forces/torques of PMSMs, there has also been method for thermal analysis of PMSMs to improve the energy-efficiency of PMSMs [51].

1.2.2 Orientation Sensing

The structural simplicity of PMSMs has motivated many researchers to seek precise manipulation of PMSMs where multi-DOF motion sensing plays an important role in order for closed-loop control. The orientation sensing of PMSMs has been achieved through a variety of techniques. In [32], a customized mechanism was designed to mechanically decouple the PMSM motion into three independent directions for measuring with three single-axis encoders/sensors. Similar schemes have also been proposed in [52] and [53]. As shown in Figs. 1.2 and 1.7, the motion-constraining mechanisms used in these methods introduce additional inertia and friction, not only limiting the PMSM bandwidth but also resulting in physical wear and tear. Inclinometers, accelerometers, and other inertia and gyroscopic sensors offer an alternative means to measure the orientation/position. However, these sensors which must be directly attached to the moving body introduce additional inertia and dynamical imbalance to the system; and additional modules/bridges are also required for power and signal transmissions. The magnetic fields of PMSMs can also affect the readings of these sensors because magnetometers are

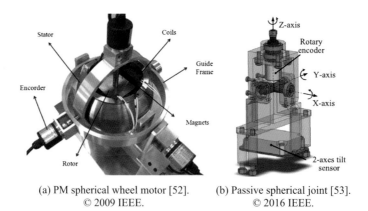

(a) PM spherical wheel motor [52]. © 2009 IEEE.

(b) Passive spherical joint [53]. © 2016 IEEE.

Fig. 1.7 Sensing with decoupling mechanisms

1.2 The State of the Art

usually incorporated in these sensors for sensor fusion in order to eliminate the drifting issues in gyroscope computation.

There are also non-contact solutions for orientation sensing of PMSMs. The image-reading system has provided a basis for developing a 3-DOF vision-based orientation sensor [54, 55] that reads encoded gridlines printed on a spherical surface (Fig. 1.8). Lee and Zhou [56] extended the vision-based method but used microscopic surface features (instead of dedicated patterns) to determine motion changes (Fig. 1.9), which led to the design concept and theory of a dual-sensor system capable of measuring 3-DOF planar and spherical motions in real time. The concept feasibility of two prototype 3-DOF dual-sensor systems for measuring the instantaneous center of rotation and the angular displacement of a moving surface was demonstrated experimentally. In [57], a vision-based approach combined with a recursive nonlinear optimization algorithm was proposed. However, the nonlinear nature of the problem requires a fairly good initial guess of the orientation. A laser-based-orientation measurement was presented in [58]; but this method required a flat reflecting plate which must be specially fabricated and mounted on the rotor.

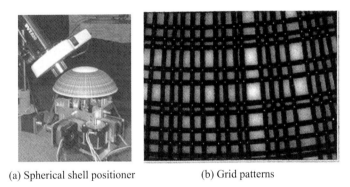

(a) Spherical shell positioner (b) Grid patterns

Fig. 1.8 Sensing with machine vision-based methods [55]

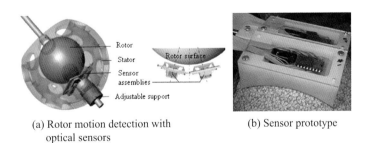

(a) Rotor motion detection with optical sensors (b) Sensor prototype

Fig. 1.9 Sensing with optical sensors [56]

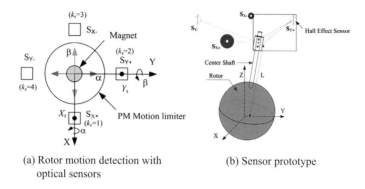

(a) Rotor motion detection with optical sensors

(b) Sensor prototype

Fig. 1.10 Magnetic field-based sensing [62]

Other than the vision and optical based methods, there have also been methods utilizing the magnetic-field measurements of the moving rotor PMs. As compared to its other non-contact counterparts, magnetic sensors do not require "a line of sight" and permit sensing across multiple non-ferromagnetic mediums. In [59], Wang, Jewell and Howe derived the 2-DOF rotor orientation in closed-form using the analytical results of the magnetic field. In [60], inverse computation of the rotor position was achieved using a nonlinear optimization algorithm to minimize the deviation between measured and modeled magnetic field (using a single dipole analytical model). This approach requires a good initial guess of the parameters and is computationally slow. A similar methodology was adopted in [61] for a decoupled multi-axis translational system. Lee and Son used the distributed multi-pole (DMP) model to characterize the magnetic field of a single PM specially installed on the rotor shaft and developed a method for 2-DOF orientation sensing using methodically placed sensors [62] (Fig. 1.10). Based on this method, an optimization strategy for orientation tracking of moving objects in 3D space was proposed in [63]. Lee and Foong [64] have developed a field-based method which used an artificial neural network (ANN) as a direct mapping for orientation determination. This method allowed the determination of the 3-DOF orientation directly from measurements of the existing magnetic field of rotor PMs. In [65] a sensing method based on back-emf was developed to detect the 3-DOF orientation and the angular velocity simultaneously by measuring the voltage changes across the EM windings. This method eliminates the need to install external sensors in PMSMs.

1.2.3 Control Methods

Motivated by the growing requirements of precise operation for multi-DOF manipulation, significant research efforts have been focused on the closed-loop orientation control of PMSMs. Lee et al. presented the dynamic model of a variable

1.2 The State of the Art

reluctance spherical motor in [32] and derived a maximum torque formula and used a look-up table based nonlinear scheme for online optimization of current inputs. They also formulated a reaction-free control strategy based on the principle of magnetic levitation in [66], and a robust back-stepping controller to compensate for imperfect modeling and computational approximations in [50]. The performance of the proposed controller was evaluated experimentally against a classical PD controller. However, the external decoupling mechanisms for orientation sensing resulting in large inertia and friction to the system and restricted the accuracy as well as the bandwidth of the control system. Wang et al. experimentally investigated a PID controller for closed-loop orientation control in 2-DOF with magnetic field sensors to detect the 2-DOF rotor orientation [67]. In [68], Son and Lee developed an operational principle to simplify motion control of a spherical wheel motor, which decouples the continuous spinning motion and the inclination of the rotor shaft, and derived a control law that combines a feed forward input-shaper for controlling the spin motion and a closed-loop PD controller for orienting the inclination of the spinning shaft. Other than classical PID controllers, modern control methods have also been applied on PMSMs to deal with the nonlinearities in rotor dynamics and the uncertainties in modeling errors and external loads. Xia, Guo and Shi applied a fuzzy controller with a neural network identifier to a spherical motor [69]; numerical investigations showed its self-adaptive ability and strong robustness against uncertainties. A similar strategy was also proposed in [70] on a spherical stepper motor. In [71], a robust adaptive iterative learning control (ILC) algorithm for 3-DOF PM spherical actuators to improve their trajectory tracking performance was presented. A new hybrid control scheme consisting of a PD feedback control with varying gains, a PD-type ILC with adjustable gains and a robust term were developed to compensate for system uncertainties. The real-time implementation of PMSM control methods usually relies on the multi-DOF orientation feedback, the sampling rate of which could degrade the effectiveness and accuracy of the controlled PMSMs. To overcome this problem, a field-feedback control (DFC) method was presented in [72], which utilizes real-time measurements of the existing rotor magnetic field in the feedback loop for controlling the multi-DOF orientation of a PMSM. As a DFC system requires only measured magnetic fields, it eliminates the need for an external orientation sensing system; and furthermore, its major components can operate independently permitting parallel processing. The DFC, which has been experimentally validated in a 3-DOF trajectory tracking control test, can greatly reduce accumulated errors and time-delay often found in other methods that rely on orientation-dependent models for feedback control of a multi-DOF PMSM.

1.3 Book Outline

Figure 1.11 schematically illustrates the design concept of a PMSM typically consisting of a socket stator (rotor) and a ball-like rotor (stator), where the rotor is concentrically supported by a ball-joint and can rotate freely in the workspace in 3-DOF. There are two typical configurations for PMSMs as illustrated in (a) and (b), where the rotor is inside the stator (contained by the socket) or outside the stator (surrounding the ball-like stator) respectively.

Generally, the former design applies for the cases when the rotor is connected with an end-effector, and the latter applies for multi-DOF stages or platforms. As both configurations operate on the working principles, the modeling, sensing and control methods introduced in this book can be applied to both designs. The motion of a PMSM rotor is a result of the electromagnetic interactions among the magnetic poles installed in the stator and rotor, which are governed by Lorentz forces of the current-carrying electromagnetic magnets (EMs) in the magnetic fields of the permanent magnets (PMs). To allow for a brushless and wire-free rotor, a moving-PM and stationary-EM configuration is preferred for PMSM designs.

Mathematically, a PMSM can be modeled as a combination of rotor dynamics and torque-current relationship. As will be illustrated in the later chapters, the equation of motion of the PMSM can be characterized by

$$[\mathbf{M}]\ddot{\mathbf{q}} + \mathbf{C}(\mathbf{q},\dot{\mathbf{q}})\dot{\mathbf{q}} + \mathbf{g}(\mathbf{q}) = \mathbf{T} \tag{1.1}$$

where $\mathbf{q}$ is the orientation; $[\mathbf{M}]$ is the inertia matrix; $\mathbf{C}(\mathbf{q},\dot{\mathbf{q}})\dot{\mathbf{q}}$ is the centripetal and Coriolis torque vector; $\mathbf{g}$ is the gravitational torque vector; and $\mathbf{T}$ is the total torque on the rotor due to the Lorentz forces of all the EMs. For illustration, the torque contributed by each EM can be expressed using the Lorentz force equation:

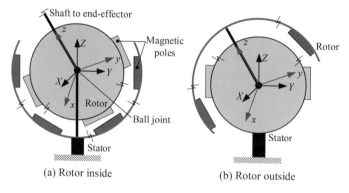

Fig. 1.11 PMSM design concepts

1.3 Book Outline

$$\mathbf{T} = -i \int_V \mathbf{r} \times (\mathbf{B} \times d\mathbf{l}) \qquad (1.2)$$

where i is the current flowing through the EM; and $d\mathbf{l}$ is the unit vector along current direction vector; $\mathbf{r}$ is the vector from the rotation center to the field point; $\mathbf{B}$ is the magnetic flux densities surrounding the EM; and V is the volume of the coil winding. As shown in Eq. (1.2), the torque at any rotor orientation depends on both the current inputs supplied into the stator EMs and the magnetic fields of the rotor PMs.

Figure 1.12 presents a block diagram of a PMSM, where the EMs are surrounded by the magnetic fields of the rotor PMs. By appropriately manipulating the current inputs in the stator EMs, the torque applied on the rotor can be controlled, which translates into the rotational motion of the rotor. The key issues in developing a PMSM are three folds: (1) Formulate the models to characterize the magnetic field, the force and torque, as well as the rotor dynamics. (2) Develop sensing techniques to detect the rotor states for feedback. (3) Derive control laws for orientation regulation, motion tracking, and/or force/torque manipulation.

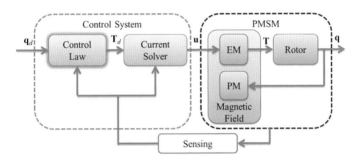

Fig. 1.12 Block diagram of a PMSM and its control system

To facilitate the PMSM studies and push forward its development for various applications, the contents of this book are divided into three major parts to elaborate the *modeling*, *sensing* and *control* of PMSMs. The chapters in this book are outlined as follows:

Part 1 (including Chaps. 2–4) presents the modeling methods for PMSMs. In Chapter 2, the analytical models characterizing the magnetic fields, force/torque, as well as the rotor dynamics, are formulated. Chapter 3 presents a distributed multi-pole (DMP) method for modeling the magnetic field and force/torque of PMSMs in the design process. With closed-form solutions, the DMP method can greatly improve the computational efficiencies. Numerical examples are provided to illustrate the DMP method for analyzing a PMSM design. In order to fulfill the requirements of real-time control, Chapter 4 presents methods for computing the magnetic force/torque in a PMSM with specified design configurations. By establishing a direct relationship among rotor orientation, torque and current inputs with kernel functions in simple forms to characterize the forces/torques in the EM/PM pole-pairs, the model-based method can be implemented with time and storage-efficient algorithms to facilitate real-time computations.

Part 2 (including Chaps. 5 and 6) presents the sensing methods for PMSMs. In Chapter 5, a magnetic field-based method is presented for orientation sensing which computes the multi-DOF orientation by measuring the existing magnetic fields of the rotor-PMs. To facilitate the adoption of modern control methods for PMSMs, Chapter 6 presents a back-emf method based on magnetic flux models to detect the motion states (including both the orientation and the angular velocities) of a PMSM.

Part 3 (including Chaps. 7 and 8) presents the control methods for PMSMs. Chapter 7 starts with an introduction of classical orientation control schemes for PMSMs, and then presents the new control method referred to here as direct field-feedback control (DFC) which bypasses the orientation sensing in control-loop and utilizes the magnetic field measurements for feedback. The method shows outstanding control accuracies with high bandwidth for trajectory tracking tests on a PMSM. Apart from motion systems, an application where the PMSM is used as a two-mode haptic device for hybrid orientation-torque manipulation is presented in Chapter 8.

References

1. K.-M. Lee, H. Son, K. Bai, Design concept and analysis of a magnetically levitated multi-dof tiltable stage for micro machining, in *International Conference on Micro-manufacturing*, 2007
2. J.K. Park, Development of next generation microfactory systems, in *2nd International Workshop on Microfactory Technology*, 2006, pp. 6–7
3. J.J. Adams, S.C. Slimmer, T.F. Malkowski, et al., Comparison of spherical antennas fabricated via conformal printing: helix, meanderline, and hybrid designs. IEEE Antennas Wirel. Propag. Lett. **10**, 1425–1428 (2011)
4. S. Yang, R.A. MacLachlan, C.N. Riviere, Manipulator design and operation of a six-degree-of-freedom handheld tremor-canceling microsurgical instrument. IEEE-ASME Trans. Mechatron. **20**, 761–772 (2015)
5. L. Chien, D.F. Chen, C.C. Lan, Design of an adaptive exoskeleton for safe robotic shoulder rehabilitation, in *2016 IEEE International Conference on Advanced Intelligent Mechatronics (AIM)*, 2016, pp. 282–287
6. Y. Wang, H. Fujimoto, S. Hara, Torque distribution-based range extension control system for longitudinal motion of electric vehicles by LTI modeling with generalized frequency variable. IEEE/ASME Trans. Mechatron. **21**, 443–452 (2016)
7. F. Petit, A. Dietrich, A. Albu-Sch, A. Albu-Schaffer, Generalizing torque control concepts: using well-established torque control methods on variable stiffness robots. IEEE Robot. Autom. Mag. **22**, 37–51 (2015)
8. K.-M. Lee, C. Kwan, Design concept development of a spherical stepper for robotic applications. IEEE Trans. Robot. Autom. **7**, 175–181 (1991)
9. B.B. Bederson, R.S. Wallace, E.L. Schwartz, A miniature pan-tilt actuator: the spherical pointing motor. IEEE Trans. Robot. Autom. **10**, 298–308 (1994)

References

10. M.K. Rashid, Z.A. Khalil, Configuration design and intelligent stepping of a spherical motor in robotic joint. J. Intell. Robot. Syst. **40**, 165–181 (2004)
11. K.-M. Lee, K. Bai, J. Lim, S. Hungsun, Spherical motor design for a multi-DOF tiltable stage, in *Proceedings of 4th International Workshop on Microfactory Technology*, 2007, pp. 31–38
12. L. Rossini, O. Chetelat, E. Onillon, Y. Perriard, Force and torque analytical models of a reaction sphere actuator based on spherical harmonic rotation and decomposition. IEEE/ASME Trans. on Mechatronics. **18**, 1006–1018 (2013).
13. K. Bai, J.J. Ji, K.M. Lee, S.Y. Zhang, A two-mode six-DOF motion system based on a ball-joint-like spherical motor for haptic applications. Comput. Math Appl. **64**, 978–987 (2012)
14. J.F.P. Fernandes, P.J.C. Branco, The shell-like spherical induction motor for low-speed traction: electromagnetic design, analysis, and experimental tests. IEEE Trans. Ind. Electron. **63**, 4325–4335 (2016)
15. H. Kawano, H. Ando, T. Hirahara, C. Yun, S. Ueha, Application of a multi-DOF ultrasonic servomotor in an auditory tele-existence robot. IEEE Trans. Robot. **21**, 790–800 (2005)
16. T. Mashimo, S. Toyama, H. Ishida, Design and implementation of spherical ultrasonic motor. IEEE Trans. Ultrason. Ferroelectr. Freq. Control **56**, 2514–2521 (2009)
17. H. Nagasawa, S. Honda, Development of a spherical motor manipulated by four wires, in *Proceedings of American Society for Precision Engineering*, Scottsdale, US, 2000, pp. 219–221
18. S.E. Wright, A.W. Mahoney, K.M. Popek, J.J. Abbott, A spherical-magnet end-effector for robotic magnetic manipulation, in *IEEE International Conference on Robotics and Automation (ICRA)*, 2015, pp. 1190–1195
19. F.C. Williams, E.R. Laithwaite, J.F. Eastham, Development and design of spherical induction motors, in *Proceedings of the IEEE—Part A: Power Engineering*, 1959, pp. 471–484
20. G. Vachtsevanos, K. Davey, K.-M. Lee, Development of a novel intelligent robotic manipulator. IEEE Control Syst. Mag. **7**, 9–15 (1987)
21. K. Davey, G. Vachtsevanos, R. Powers, The analysis of fields and torques in spherical induction motors. IEEE Trans. Magn. **23**, 273–282 (1987)
22. A. Foggia, E. Olivier, F. Chappuis, J. C. Sabonnadiere, A new three degrees of freedom electromagnetic actuator, in *Industry Applications Society Annual Meeting*, 1988, pp. 137–141
23. B. Dehez, G. Galary, D. Grenier, B. Raucent, Development of a spherical induction motor with two degrees of freedom. IEEE Trans. Magn. **42**, 2077–2089 (2006)
24. M. Kumagai, R.L. Hollis, Development and control of a three DOF spherical induction motor, in *2013 IEEE International Conference on Robotics and Automation*, 2013, pp. 1528–1533
25. J.F.P. Fernandes, P.J.C. Branco, The shell-like spherical induction motor for low-speed traction: electromagnetic design, analysis, and experimental tests. IEEE Trans. Ind. Electron. **63**, 4325–4335 (2016)
26. R.L. Hollis, S.E. Salcudean, A.P. Allan, A six-degree-of-freedom magnetically levitated variable compliance fine motion wrist: Design, modeling and control. IEEE Trans. Robot. Autom **7**, 320–332 (1991)
27. K. Kaneko, I. Yamada, K. Itao, A spherical DC servo motor with three degrees of freedom. ASME Dyn. Sys. Control Div, 398–402 (1988)
28. K.-M. Lee, G. Vachtsevanos, C. Kwan, Development of a spherical stepper wrist motor, in *Proceedings of 1988 IEEE International Conference on Robotics and Automation*, Vol. 1, pp. 267–272, 1988
29. K.-M. Lee, C. Kwan, Design concept development of a spherical stepper for robotic applications. IEEE Trans. Robot. Autom. **7**, 175–181 (1991)
30. G.S. Chirikjian, D. Stein, Kinematic design and commutation of a spherical stepper motor. IEEE/ASME Trans. Mechatron. **4**, 342–353 (1999)
31. K.-M. Lee, J. Pei, R. Roth, Kinematic analysis of a three degree-of-freedom spherical wrist actuator. Mechatronics **4**, 581–605 (1994)

32. K.-M. Lee, R. Roth, Z. Zhou, Dynamic modeling and control of a ball-joint-like variable-reluctance spherical motor. J. Dyn. Syst. Meas. Control **118**, 29–40 (1996)
33. H. Son, K.-M. Lee, Distributed multipole models for design and control of PM actuators and sensors. IEEE-ASME Trans. Mechatron. **13**(2), 228–238 (2008)
34. H. Son, K.-M. Lee, Open-loop controller design and dynamic characteristics of a spherical wheel motor. IEEE Trans. Ind. Electron. **57**, 3475–3482 (2010)
35. L. Yan, I.M. Chen, G. Yang, K.-M. Lee, Analytical and experimental investigation on the magnetic field and torque of a permanent magnet spherical actuator. IEEE/ASME Trans. Mechatron. **11**(4), 409–419 (2006)
36. L. Yan, I.M. Chen, C.K. Lim, G. Yang, W. Lin, K.-M. Lee, Design and analysis of a permanent magnet spherical actuator. IEEE/ASME Trans. Mechatron. **13**, 239–248 (2008)
37. L. Yan, I.-M. Chen, C.K. Lim, G. Yang, W. Lin, K.-M. Lee, Hybrid torque modeling of spherical actuators with cylindrical-shaped magnet poles, (in English). Mechatronics **21**(1), 85–91 (2011)
38. L. Yan, I.-M. Chen, C.K. Lim, G. Yang, K.-M. Lee, *Design, modeling and experiments of 3-DOF electromagnetic spherical actuators* (Springer, Reading, MA, 2011)
39. H.Y. Kim, H. Kim, D.G. Gweon, J. Jeong, Development of a novel spherical actuator with two degrees of freedom. IEEE/ASME Trans. Mechatron. **20**, 532–540 (2015)
40. L. Rossini, E. Onillon, O. Chételat, Y. Perriard, Closed-loop magnetic bearing and angular velocity control of a reaction sphere actuator. Mechatronics **30**, 214–224 (2015)
41. L. Rossini, S. Mingard, A. Boletis, E. Forzani, E. Onillon, Y. Perriard, Rotor design optimization for a reaction sphere actuator. IEEE Trans. Ind. Appl. **50**, 1706–1716 (2014)
42. K. Kahlen, I. Voss, C. Priebe, R.W. Doncker, Torque control of a spherical machine with variable pole pitch. IEEE Trans. Power Electron. **19**(7), 1628–1634 (2004)
43. K.-M. Lee, H. Son, Distributed multipole model for design of permanent-magnet-based actuators. IEEE Trans. Magn. **43**(10), 3904–3913 (2007)
44. K.M. Lee, K. Bai, J. Lim, Dipole models for forward/inverse torque computation of a spherical motor. IEEE/ASME Trans. on Mechatronics. **14**(1), 46–54 (2009)
45. H. Son, K. Bai, J. Lim, K.M. Lee, Design of multi-DOF electromagnetic actuators using distributed multipole models and image method. Int. J. Appl. Electromagnet. Mech. **34**(3), 195–210 (2010)
46. J. Lim, K.M. Lee, Distributed multilevel current models for design analysis of electromagnetic actuators. IEEE/ASME Trans. Mechatron. **20**(5), 2413–2424 (2015)
47. J. Lim, K.M. Lee, Design of electromagnetic actuators using layout optimization with distributed current source models. IEEE/ASME Trans. Mechatron. **20**(6), 2726–2735 (2015)
48. K.-M. Lee, R. Sosseh, Z. Wei, Effects of the torque model on the control of a VR spherical motor. Control Eng. Pract. **12**, 1437–1449 (2004)
49. Z. Qian, Q. Wang, L. Ju, A. Wang, J. Liu, Torque modeling and control algorithm of a permanent magnetic spherical motor, in *International Conference on Electrical Machines and Systems*, 2009, pp. 1–6
50. K.-M. Lee, H. Son, Torque model for design and control of a spherical wheel motor, in *Proceedings of the 2005 IEEE/ASME International Conference on Advanced Intelligent Mechatronics*, 2005, pp. 335–340
51. H. Li, Y. Shen, Thermal analysis of the permanent-magnet spherical motor. IEEE Trans. Energy Convers. **30**, 991–998 (2015)
52. D.W. Kang, W.H. Kim, S.C. Go, C.S. Jin, S.H. Won, D.H. Koo, J. Lee, Method of current compensation for reducing error of holding torque of permanent-magnet spherical wheel motor. IEEE Trans. Magn. **45**, 2819–2822 (2009)
53. L. Zhang, W. Chen, J. Liu, C. Wen, A robust adaptive iterative learning control for trajectory tracking of permanent-magnet spherical actuator. IEEE Trans. Ind. Electron. **63**, 291–301 (2016)
54. K.-M. Lee, G. Meyouhas, R. Blenis, A machine-vision-based wrist sensor for direct measurement of three-DOF orientation. Mechatronics **3**(5), 571–587 (1993)

References

55. H. Garner, M. Klement, L. Kok-Meng, Design and analysis of an absolute non-contact orientation sensor for wrist motion control, in *IEEE/ASME International Conference on Advanced Intelligent Mechatronics,* Vol. 1, 2001, pp. 69–74
56. K.-M. Lee, D. Zhou, A real-time optical sensor for simultaneous measurement of three-DOF motions. IEEE/ASME Trans. Mechatron. **9**, 499–507 (2004)
57. D. Stein, E.R. Scheinerman, G.S. Chirikjian, Mathematical models of binary spherical-motion encoders. IEEE/ASME Trans. Mechatron. **8**(2), 234–244 (2003)
58. L. Yan, I.-M. Chen, Z. Guo, Y. Lang, Y. Li, A three degree-of-freedom optical orientation measurement method for spherical actuator applications. IEEE Trans. Autom. Sci. Eng. **8**(2), 319–326 (2011)
59. J. Wang, G.W. Jewell, D. Howe, A novel spherical actuator: design and control. IEEE Trans. Magn. **33**, 4209–4211 (1997)
60. C. Hu, M.Q.-H. Meng, M. Mandal, A linear algorithm for tracing magnet position and orientation by using three-axis magnetic sensors. IEEE Trans. Magn. **47**, 4096–4101 (2007)
61. M. Tsai, C. Yang, A flux-density-based electromagnetic servo system for real-time magnetic servoing/tracking. IEEE/ASME Trans. on Mechatronics. **13**, 249–256 (2008)
62. H. Son, K.-M. Lee, Two-DOF magnetic orientation sensor using distributed multipole models for spherical wheel motor. Mechatronics **21**(1), 156–165 (2011)
63. W. Fang, H. Son, Optimization of measuring magnetic fields for position and orientation tracking. IEEE/ASME Trans. on Mechatronics. **16**(3), 440–448 (2011)
64. S. Foong, K.-M. Lee, K. Bai, Magnetic field-based sensing method for spherical joint, in *Proceedings IEEE International Conference on Robotics and Automation*, (Anchorage, AK, 2010), pp. 5447–5452
65. K. Bai, K.-M. Lee, J. Lu, A magnetic flux model based method for detecting multi-DOF motion of a permanent magnet spherical motor. Mechatronics **39**, 217–225 (2016)
66. Z. Zhi, L. Kok-Meng, Real-time motion control of a multi-degree-of-freedom variable reluctance spherical motor, in *Proceedings of IEEE International Conference on Robotics and Automation*, Vol. 3, 1996, pp. 2859–2864
67. J. Wang, G.W. Jewell, D. Howe, A novel spherical actuator: design and control. Magn. IEEE Trans. on **33**, 4209–4211 (1997)
68. H. Son, K.-M. Lee, Control system design and input shape for orientation of spherical wheel motor. Control Eng. Pract. **24**, 120–128 (2014)
69. C. Xia, C. Guo, T. Shi, A neural-network-identifier and fuzzy-controller-based algorithm for dynamic decoupling control of permanent-magnet spherical motor. IEEE Trans. Ind. Electron. **57**, 2868–2878 (2010)
70. L. Zheng, W. Qunjing, Robust neural network controller design for permanent magnet spherical stepper motor, in *ICIT 2008. IEEE International Conference on Industrial Technology*, 2008, pp. 1–6
71. L. Zhang, W. Chen, J. Liu, C. Wen, A robust adaptive iterative learning control for trajectory tracking of permanent-magnet spherical actuator. IEEE Trans. Ind. Electron. **63**, 291–301 (2016)
72. K. Bai, K.M. Lee, Direct field-feedback control of a ball-joint-like permanent-magnet spherical motor. IEEE/ASME Trans. Mechatron. **19**, 975–986 (2014)

Part I
Modeling Methods

Chapter 2
General Formulation of PMSMs

As an energy-converting system, a permanent magnet spherical motor (PMSM) can be characterized as a combination of two subsystems; an electromagnetic system and a rotor dynamic system. The analytical models presented here include the equations governing the magnetic field, energy and force/torque of a PMSM, and the equations of motion describing the three degree-of-freedom (DOF) rotor dynamics.

2.1 PMSM Electromagnetic System Modeling

2.1.1 Governing Equations of Electromagnetic Field

The three-DOF motion of spherical motors is a direct result of the electromagnetic interactions among the magnetic fields of the stator electromagnets (EMs) and the rotor permanent magnets (PMs) as illustrated in Fig. 2.1. The electromagnetics involved are governed by the following two Maxwell's equations for a quasi-static[1] regime, where **H** is the magnetic field intensity; **B** is the magnetic flux density; **J** is the volume density of free current:

$$\nabla \times \mathbf{H} = \mathbf{J} \quad (2.1)$$

$$\nabla \cdot \mathbf{B} = 0 \quad (2.2)$$

When magnetic materials are involved, the relationship between **B** and **H** can be characterized by (2.3) where μ_0 is the permeability of free space; **M** is the material magnetization vector:

---

[1] The travel time of light can be neglected and the fields are as if they propagated instantaneously [1].

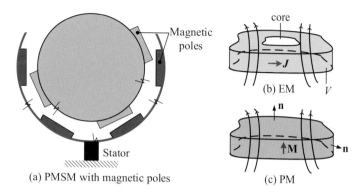

Fig. 2.1 Illustrations of magnetic poles and EM/PM magnetic fields

$$\mathbf{B} = \mu_0 \mathbf{H} + \mathbf{M} \qquad (2.3)$$

When the magnetic properties of the medium are linear and isotropic, the magnetization is directly proportional to **H**, (2.3) can be also expressed as (2.3a):

$$\mathbf{B} = \mu \mathbf{H} \qquad (2.3a)$$

where μ denotes the absolute permeability of the medium.

Assuming no magnetic saturation in the materials, (2.1)–(2.3) lead to specific linear partial differential equations (PDEs), which can be solved to characterize the magnetic fields for a given electromagnetic motor configuration. As the sum of any particular solutions of a linear differential equation is also a solution to the equation, the governing equations along with their particular solutions in free space are derived for the following special cases:

Special Case 1: Magnetic Material Free and Biot-Savart Law: **M** = 0, **J** is given

Since **B** is divergence free, a vector magnetic potential **A** is introduced such that

$$\mathbf{B} = \nabla \times \mathbf{A} \qquad (2.4)$$

Substituting (2.4) into (2.1) yields:

$$\nabla(\nabla \cdot \mathbf{A}) - \nabla^2 \mathbf{A} = \mu_0 \mathbf{J} \qquad (2.5)$$

By choosing $\nabla \cdot \mathbf{A} = 0$ (Coulomb gauge [1]), the governing Eq. (2.5) becomes

$$\nabla^2 \mathbf{A} = -\mu_0 \mathbf{J} \qquad (2.6)$$

2.1 PMSM Electromagnetic System Modeling

This is a vector Poisson's equation and the solution can be obtained as [2]:

$$\mathbf{A} = \frac{\mu_0}{4\pi} \int_V \frac{\mathbf{J}}{|\mathbf{R} - \mathbf{R}'|} dv \tag{2.7}$$

In (2.7), $|\mathbf{R} - \mathbf{R}'|$ is the vector from the source point $\mathbf{R}'$ to the field point $\mathbf{R}$; V is the volume of the conductor carrying free current (as shown in Fig. 2.1b). The magnetic flux density $\mathbf{B}$ can be obtained from (2.4) as the Biot-Savart law:

$$\mathbf{B} = \frac{\mu_0}{4\pi} \int_V \frac{\mathbf{J} \times (\mathbf{R} - \mathbf{R}')}{|\mathbf{R} - \mathbf{R}'|^3} dv \tag{2.8}$$

(2.8) is often used to compute the magnetic flux densities of the electromagnets (EMs), as shown in Fig. 2.1b.

Special Case 2: Current Free: $\mathbf{J} = 0$, $\mathbf{M}$ *is given*

When there is no free current in the space, (2.1) becomes $\nabla \times \mathbf{H} = 0$, a magnetic scalar potential Φ is introduced here such that

$$\mathbf{H} = -\nabla \Phi \tag{2.9}$$

Substituting (2.9) and (2.3) into (2.2), the governing equation for this case can be obtained as:

$$\nabla^2 \Phi = \nabla \cdot \mathbf{M} \tag{2.10}$$

The solution to the Poisson's equation can be obtained as:

$$\Phi = -\frac{1}{4\pi} \int_V \frac{\nabla \cdot \mathbf{M}}{|\mathbf{R} - \mathbf{R}'|} dv + \frac{1}{4\pi} \int_S \frac{\mathbf{M} \cdot \mathbf{n}}{|\mathbf{R} - \mathbf{R}'|} ds \tag{2.11}$$

where V is the volume of the magnetized material (magnetization $\mathbf{M}$); and S is the surface bounding V; $\mathbf{n}$ is the surface normal of S, as shown in Fig. 2.1c. In particular, if the region out of V is free space, $\mathbf{B}$ can be obtained as (2.12) based on Φ expressed in (2.11):

$$\mathbf{B} = \frac{\mu_0}{4\pi} \int_V \frac{-(\nabla \cdot \mathbf{M})(\mathbf{R} - \mathbf{R}')}{|\mathbf{R} - \mathbf{R}'|^3} dv + \frac{\mu_0}{4\pi} \int_S \frac{(\mathbf{M} \cdot \mathbf{n})(\mathbf{R} - \mathbf{R}')}{|\mathbf{R} - \mathbf{R}'|^3} ds \tag{2.12}$$

Similarly, if we choose to write $\mathbf{B} = \nabla \times \mathbf{A}$ (with Coulomb gauge $\nabla \cdot \mathbf{A} = 0$) to satisfy $\nabla \cdot \mathbf{B} = 0$ automatically, (2.1) and (2.3) will yield

$$\nabla \times \mathbf{H} = \nabla \times (\mathbf{B}/\mu_0 - \mathbf{M}) = 0 \qquad (2.13)$$

The governing equation can be derived in the form of a Poisson's equation:

$$\nabla^2 \mathbf{A} = -\mu_0 \nabla \times \mathbf{M} \qquad (2.14)$$

The solution to $\mathbf{A}$ can be obtained as (2.15) in free space:

$$\mathbf{A} = \frac{\mu_0}{4\pi} \int_V \frac{\nabla \times \mathbf{M}}{|\mathbf{R} - \mathbf{R}'|} dv + \frac{\mu_0}{4\pi} \int_S \frac{\mathbf{M} \times \mathbf{n}}{|\mathbf{R} - \mathbf{R}'|} ds \qquad (2.15)$$

(2.11), (2.12) and (2.15) are often used to compute the scalar magnetic potential, the magnetic flux densities and the vector magnetic potential of permanent magnets (PMs) respectively in free space, as shown in Fig. 2.1c.

Special Case 3: Magnetic Material and Current Free: $\mathbf{J} = 0$, $\mathbf{M} = 0$

If both current density and magnetizations vanish in some finite region of space, with (2.1) becoming $\nabla \times \mathbf{H} = 0$, the governing equation can be obtained as (2.16) in a Laplacian form based on (2.2) and (2.3):

$$\nabla^2 \Phi = 0 \qquad (2.16)$$

The solutions to the Laplacian equation can be solved using separation of variables with specified boundary conditions. For spherical motor analysis, in particular, the Laplace equation can be expressed as (2.17) in spherical coordinates:

$$\frac{1}{r^2}\frac{\partial^2}{\partial r^2}(r\Phi) + \frac{1}{r^2 \sin\theta}\frac{\partial}{\partial \theta}\left(\sin\theta \frac{\partial \Phi}{\partial \theta}\right) + \frac{1}{r^2 \sin\theta}\left(\sin\theta \frac{\partial^2 \Phi}{\partial \phi^2}\right) = 0 \qquad (2.17)$$

where (r, θ, ϕ) are the radial distance, polar and azimuthal angles.

2.1.2 Boundary Conditions

At the interface between two neighboring regions with different permeability, the following two boundary conditions must be satisfied:

Condition I: The normal component of $\mathbf{B}$ *is continuous across the boundary.*

$$(\mathbf{B}_1 - \mathbf{B}_2) \cdot \mathbf{n} = 0 \qquad (2.18)$$

2.1 PMSM Electromagnetic System Modeling

Condition II: The tangential component of **H** *is continuous along the boundary.*

$$(\mathbf{H}_1 - \mathbf{H}_2) \cdot \mathbf{t} = 0 \quad \text{or} \quad (\mathbf{H}_1 - \mathbf{H}_2) \times \mathbf{n} = 0 \tag{2.19}$$

In (2.18) and (2.19), the subscripts "1" and "2" denote the regions across the boundary respectively; **n** and **t** represent the normal and tangent vectors across the boundary.

2.1.3 Magnetic Flux Linkage and Energy

The instant magnetic energy stored in an electromagnetic system can be written in terms of the magnetic fields as

$$W = \int_V (\mathbf{H} \cdot \mathbf{B}) dv \tag{2.20}$$

The magnetic energy can be also expressed in terms of current density and magnetic vector potential when there is free current in the system:

$$W = \int_V (\mathbf{A} \cdot \mathbf{J}) dv \tag{2.21}$$

Note that the volume V should be sufficiently large to contain the interested object.

In most actuator systems including PMSMs, the current-carrying conductors are in the form of a winding coil consisting of contiguous wires where the supplied current with uniform current density can be assumed. Consider Fig. 2.2a where a closed-loop wire conductor C is surrounded by an external magnetic field **B**. The magnetic flux through the surface S which is enclosed by C can be obtained as:

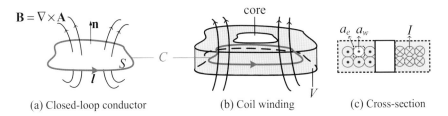

(a) Closed-loop conductor   (b) Coil winding   (c) Cross-section

Fig. 2.2 Illustrations of a winding coil

$$\Psi_C = \int_S (\mathbf{B} \cdot \mathbf{n}) ds \tag{2.22}$$

where **n** represents the surface normal of S. With the magnetic vector potential **A** defined in (2.4) and the Stokes' theorem, the surface integral (2.22) can be reduced to a line integral (2.23) where l is the directional vector of C:

$$\Psi_C = \int_C (\mathbf{A} \cdot \mathbf{l}) dc \tag{2.23}$$

When the wire is supplied with current I, the magnetic energy of the loop is

$$W_C = I\Psi_C \tag{2.24}$$

For a coil which consists of contiguous wire conductors as shown in Fig. 2.2b, the total magnetic flux linkage (MFL) Λ through the EM can be obtained by extending (2.23) to a volume integral (2.25a) and the total magnetic energy can be obtained as (2.25b):

$$\Lambda = \frac{1}{a_e} \int_V (\mathbf{A} \cdot \mathbf{l}) dv; \quad W = I\Lambda \tag{2.25a, b}$$

In (2.25a), a_e is the equivalent cross-sectional area of the wire (as shown in Fig. 2.2c). Note that **B** and **A** in (2.22), (2.23) and (2.25a) refer to the external magnetic field excluding that contributed by the coil itself.

2.1.4 Magnetic Force/Torque

The magnetic force/torque can be computed from the following two methods Lorentz force and energy method (virtual work), while the latter can be expressed in terms of Maxwell stress tensor and flux linkage.

Lorentz Force Equation

Lorentz force equation is commonly used to calculate the magnetic force or torque exerted on current-carrying conductors surrounded by an external magnetic field **B**:

$$\mathbf{F} = \int_V \mathbf{J} \times \mathbf{B} dv; \quad \mathbf{T} = \int_V \mathbf{r} \times (\mathbf{J} \times \mathbf{B}) dv \tag{2.26a, b}$$

2.1 PMSM Electromagnetic System Modeling

In (2.26b), **r** is the moment arm from the rotation center to the source point.

Maxwell Stress Tensor

Alternatively, the magnetic force can be derived using the principle of virtual work:

$$\mathbf{F}(\text{or } \mathbf{T}) = \nabla W \tag{2.27}$$

If the gradient is taken with respect to linear variables, (2.27) leads to magnetic forces; if the gradient is taken with respect to angular variables, (2.27) leads to magnetic torques. By substituting (2.20) into (2.27), the magnetic force can be written in terms of Maxwell stress tensor as

$$\mathbf{F} = \oint_S \mathbf{\Gamma} ds \quad \text{where} \quad \mathbf{\Gamma} = \frac{1}{\mu_0}\left(\mathbf{B}(\mathbf{B}\cdot\mathbf{n}) - \frac{1}{2}|\mathbf{B}|^2\mathbf{n}\right) \tag{2.28}$$

where S is an arbitrary boundary surface enclosing the body of interest; and **n** is the normal of S. **B** is the total magnetic flux densities at S.

Permeance-Based Model

For winding coils, the magnetic force can also be obtained based on the virtual work and the magnetic flux linkage (or permeance) by substituting (2.25b) into (2.27):

$$\mathbf{F}(\text{or } \mathbf{T}) = I\nabla \Lambda \tag{2.29}$$

2.2 PMSM Rotor Dynamics

Consider a spherical motor with its rotor supported by a ball-joint on the stator as illustrated in Fig. 2.3a where XYZ (stationary) and xyz (moving) represent the stator and the rotor coordinate frames respectively. The rotor rotates freely about the ball-joint with continuous 3-DOF motion. The commonly used representations for the rotor orientation include xyz Euler angles (roll-pitch-yaw) and zyz Euler angles, as shown in Fig. 2.3b, c.

Assuming the rotor is axis-symmetric and that the center of mass coincides with the rotation center, the dynamic model of a spherical motor is derived using

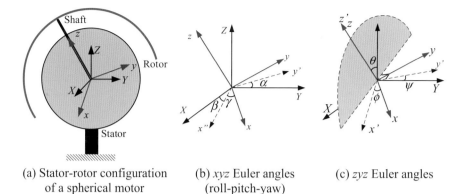

(a) Stator-rotor configuration of a spherical motor

(b) xyz Euler angles (roll-pitch-yaw)

(c) zyz Euler angles

Fig. 2.3 Illustration of coordinate systems of spherical motors ($x'y'z'$ and $x''y''z''$ are intermediate axes)

Lagrange formulation with two different representations of rotor orientation **q** in the following discussion.

(a) *xyz* Euler angles, $\mathbf{q} = (\alpha, \beta, \gamma)^T$, Fig. 2.3b

For any orientation, the rotation matrix from stator coordinate XYZ (with unit vectors $\vec{I}, \vec{J}, \vec{K}$) to rotor frame xyz (with unit vectors $\vec{i}, \vec{j}, \vec{k}$) as illustrated in Fig. 2.3b can be obtained as:

$$\left\{\begin{array}{c} \vec{i} \\ \vec{j} \\ \vec{k} \end{array}\right\} = [\mathbf{R}] \left\{\begin{array}{c} \vec{I} \\ \vec{J} \\ \vec{K} \end{array}\right\} \tag{2.30a}$$

where

$$[\mathbf{R}] = [\mathbf{R}_\gamma][\mathbf{R}_\beta][\mathbf{R}_\alpha] = \begin{bmatrix} C_\gamma C_\beta & S_\gamma C_\alpha + C_\gamma S_\beta S_\alpha & S_\gamma S_\alpha - C_\gamma S_\beta C_\alpha \\ -S_\gamma C_\beta & C_\gamma C_\alpha - S_\gamma S_\beta S_\alpha & C_\gamma S_\alpha + S_\gamma S_\beta C_\alpha \\ S_\beta & -C_\beta S_\alpha & C_\beta C_\alpha \end{bmatrix} \tag{2.30b}$$

and

$$[\mathbf{R}_\alpha] = \begin{bmatrix} 1 & 0 & 0 \\ 0 & C_\alpha & S_\alpha \\ 0 & -S_\alpha & C_\alpha \end{bmatrix}, \quad [\mathbf{R}_\beta] = \begin{bmatrix} C_\beta & 0 & -S_\beta \\ 0 & 1 & 0 \\ S_\beta & 0 & C_\beta \end{bmatrix}, \quad [\mathbf{R}_\gamma] = \begin{bmatrix} C_\gamma & S_\gamma & 0 \\ -S_\gamma & C_\gamma & 0 \\ 0 & 0 & 1 \end{bmatrix}$$
$$(2.30c, d, e)$$

In (2.30b)–(2.30e), S and C represent sine and cosine respectively. The angular velocity of the rotor can be expressed as

2.2 PMSM Rotor Dynamics

$$\vec{\omega} = \dot{\alpha}\vec{I} + \dot{\beta}\vec{j'} + \dot{\gamma}\vec{k} = \left(\dot{\alpha}C_\gamma C_\beta + \dot{\beta}S_\gamma\right)\vec{i} + \left(-\dot{\alpha}S_\gamma C_\beta + \dot{\beta}C_\gamma\right)\vec{j} + \left(\dot{\alpha}S_\beta + \dot{\gamma}\right)\vec{k} \quad (2.31)$$

where $\vec{j'}$ represent the unit vector of y' axis of the intermediate frame as shown in Fig. 2.3b. The kinetic energy is given by (2.32) in terms of the inertia matrix $[\mathbf{I}]$ in (2.33) where $I_a = I_{zz}$ and $I_t = I_{xx} = I_{yy}$:

$$T_{KE} = \left(\frac{1}{2}\vec{\omega}^T I \vec{\omega}\right) = \frac{1}{2}\left(I_t\dot{\alpha}^2 C_\beta^2 + I_t\dot{\beta}^2 + I_a\dot{\alpha}^2 S_\beta^2 + I_a\dot{\gamma}^2 + 2I_a\dot{\alpha}\dot{\gamma}S_\beta\right) \quad (2.32)$$

$$[\mathbf{I}] = \begin{bmatrix} I_t & 0 & 0 \\ 0 & I_t & 0 \\ 0 & 0 & I_a \end{bmatrix} \quad (2.33)$$

The virtual displacement vector can be represented using

$$\begin{aligned}\delta\vec{r} &= \delta\alpha\vec{I} + \delta\beta\vec{j'} + \delta\gamma\vec{k} \\ &= \left(\delta\alpha C_\gamma C_\beta + \delta\beta S_\gamma\right)\vec{i} + \left(\delta\beta C_\gamma - \delta\alpha S_\gamma C_\beta\right)\vec{j} + \left(\delta\alpha S_\beta + \delta\gamma\right)\vec{k}\end{aligned} \quad (2.34)$$

Thus the generalized force can be derived using

$$\mathbf{Q} = \mathbf{T}\cdot\delta\vec{r} = \left(T_x C_\gamma C_\beta - T_y S_\gamma C_\beta + T_z S_\beta\right)\delta\alpha + \left(TS_\gamma + T_y C_\gamma\right)\delta\beta + T_z\delta\gamma \quad (2.35)$$

where T_x, T_y, T_z are the components of the total electromagnetic torque applied on the rotor with respect to the rotor frame. Based on (2.32) and (2.35), the Lagrange formulation has the form:

$$\frac{d}{dt}\left(\frac{\partial T_{KE}}{\partial \dot{q}_i}\right) - \frac{\partial T_{KE}}{\partial q_i} = Q_i \quad (2.36)$$

In (2.36), $i = (1, 2, 3)$; $(q_1, q_2, q_3) = (\alpha, \beta, \gamma)$; and Q_1, Q_2, Q_3 are the generalized forces which equal to the coefficients of $\delta\alpha$, $\delta\beta$, $\delta\gamma$ in (2.35). Therefore, the equations of motion derived using (2.36) have the form:

$$[\mathbf{M}]\ddot{\mathbf{q}} + \mathbf{C}(\mathbf{q}, \dot{\mathbf{q}}) = \mathbf{T} \quad (2.37)$$

where

$$[\mathbf{M}] = \begin{bmatrix} I_t C_\beta^2 + I_a S_\beta^2 & 0 & I_a S_\beta \\ 0 & I_t & 0 \\ I_a S_\beta & 0 & I_a \end{bmatrix}, \quad (2.37a)$$

$$\mathbf{C}(\mathbf{q},\dot{\mathbf{q}}) = \begin{bmatrix} 2I_a\dot{\alpha}\dot{\beta}S_\beta C_\beta + I_a\dot{\beta}\dot{\gamma}C_\beta - 2I_t\dot{\alpha}\dot{\beta}C_\beta S_\beta \\ I_t\dot{\alpha}^2 C_\beta S_\beta - I_a\dot{\alpha}^2 C_\beta S_\beta - I_a\dot{\alpha}\dot{\gamma}C_\beta \\ I_a\dot{\alpha}\dot{\beta}C_\beta \end{bmatrix} \quad (2.37b)$$

and

$$\mathbf{T} = \begin{bmatrix} C_\gamma C_\beta & -S_\gamma C_\beta & S_\beta \\ S_\gamma & C_\gamma & 0 \\ 0 & 0 & 1 \end{bmatrix} \begin{bmatrix} T_x \\ T_y \\ T_z \end{bmatrix} \quad (2.37c)$$

(b) *zyz* Euler angles, $\mathbf{q} = [\psi \quad \theta \quad \phi]^T$, Fig. 2.3c

The dynamic model can be formulated in a similar manner as the form:

$$[\mathbf{M}]\ddot{\mathbf{q}} + \mathbf{C}(\mathbf{q},\dot{\mathbf{q}})\dot{\mathbf{q}} = \mathbf{T} \quad (2.38)$$

where

$$[\mathbf{M}] = \begin{bmatrix} I_t C_\beta^2 + I_a S_\beta^2 & 0 & I_a S_\beta \\ 0 & I_t & 0 \\ I_a S_\beta & 0 & I_a \end{bmatrix}, \quad (2.38a)$$

$$\mathbf{C}(\mathbf{q},\dot{\mathbf{q}}) = \begin{bmatrix} 2\dot{\psi}S_\theta C_\theta \dot{\theta}(I_t - I_a) - S_\theta \dot{\theta}\dot{\phi}I_a \\ \dot{\psi}^2 S_\theta C_\theta (I_a - I_t) + \dot{\psi}S_\theta \dot{\phi}I_a \\ -I_a\dot{\psi}\dot{\theta}S_\theta \end{bmatrix} \quad (2.38b)$$

and

$$\mathbf{T} = \begin{bmatrix} -S_\theta C_\phi & S_\theta S_\phi & C_\theta \\ S_\phi & C_\phi & 0 \\ 0 & 0 & 1 \end{bmatrix} \begin{bmatrix} T_x \\ T_y \\ T_z \end{bmatrix} \quad (2.38c)$$

References

1. J. Jackson, *Classical Electrodynamics* (Wiley, Hoboken, 1998)
2. D.K. Cheng, *Field and Wave Electromagnetics* (Pearson, London, 1989)

Chapter 3
Distributed Multi-pole Models

The design and analysis of a PMSM require a good understanding of the magnetic fields, forces and torques, which are computationally demanding. Conventional methods for analyzing magnetic fields, forces and torques involve surface or volume integrals; thus, it takes enormous computational time during the design and analysis processes. The interests to develop a computational-efficient method for analyzing and controlling an electromagnetic system have led to the distributed multi-pole (DMP) method [1, 2] that allows for closed-form computations of its magnetic field, force and torque. The method not only takes into accounts the PM/EM geometries but also characterizes the boundary condition problems with neat forms that greatly ease the design and analysis of a PMSM.

3.1 Distributed Multi-pole Model for PMs

The magnetic field intensity **H** of a source +m (or a sink −m) can be expressed [3] in spherical coordinates (R, θ, ϕ) defined in Fig. 3.1 as

$$H_r = \frac{(-1)^j}{4\pi R^2}\left[m(t-R/c) + \frac{R}{c}\frac{\partial m(t-R/c)}{\partial t}\right]; \quad H_\theta = H_\phi = 0 \quad (3.1)$$

where c is the speed of light; m is the strength of the pole; and j takes the value 0 or 1 designating that the pole is a source or a sink respectively.

Consider PM-based actuator applications where the order of the length scale is relatively small, and consequently R/c is several orders smaller than that of the actuator characteristic time. For a current-free region (**J** = 0) where the medium is homogeneous, the field is irrotational ($\nabla \times \mathbf{B} = 0$). The magnetic flux density **B** is curl-free and can be expressed as the gradient of a scalar magnetic potential Φ:

Fig. 3.1 Field intensity of a source m in spherical coordinates

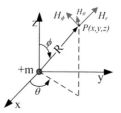

$$\mathbf{H} = -\nabla\Phi; \quad \mathbf{B} = -\mu_0 \mathbf{H} \tag{3.2}$$

where μ_0 is the permeability of free space. Since the field is continuous ($\nabla \bullet \mathbf{B} = 0$),

$$\nabla^2 \Phi(\mathbf{R}, t) = 0 \tag{3.3}$$

Except at the source or the sink, the solutions to Laplace's Eq. (3.3) satisfying the field **H** in (3.1) are given by (3.4):

$$\Phi = \frac{(-1)^j}{4\pi R} m(t) \tag{3.4}$$

Equation (3.4) can be used to derive approximate flux paths of a magnet. Two models (pole and doublet) were suggested in [4]. The *pole model* uses a source-and-sink pair at the ends of the magnet. However, as the fields of a physical magnet are everywhere finite, the poles (source and sink) attached at the end of the magnet are singularities (or infinite field density) resulting in significant errors. This is particularly critical at the air gap that is often very small as compared to other dimensions. The *doublet model* uses a single dipole at the origin of the magnet and thus, the magnetic fields outside of the physical magnet being approximated are generally finite. However, the doublet model (much like the pole model) cannot account for the effects of the shape and property of the magnet. An alternative method is to use multiple dipoles to account for the shape of the physical magnet. This method bases on the fact that Laplace equation is linear and thus the principle of superposition is applicable; in other words, the magnetic field of a PM can be characterized by the sum of the magnetic fields contributed by an appropriate distribution of sources and sinks.

3.1.1 PM Field with DMP Model

A *dipole* is defined here as a pair of source and sink separated by a distance $\bar{\ell}$. A general DMP model with k loops (or columns) of n dipoles can be derived as follows. The potential $\Phi(x, y, z)$ at any point $P(x, y, z)$ contributed by all the dipoles (in terms of the ith dipole in the jth loop) is thus given by

3.1 Distributed Multi-pole Model for PMs

$$\Phi = \sum_{j=0}^{k}\sum_{i=0}^{n} m_{ji}\varphi_{ji} = \underline{\varphi}^T \underline{m} \tag{3.5}$$

where

$$\underline{\varphi}^T = [(\varphi_{00} \cdots \varphi_{0n})\ (\varphi_{10} \cdots \varphi_{1n})\ (\cdots)\ (\varphi_{k0} \cdots \varphi_{kn})];$$

$$\underline{m} = [(m_{00} \cdots m_{0n})\ (m_{10} \cdots m_{1n})\ (\cdots)\ (m_{k0} \cdots m_{kn})]^T;$$

$$\varphi_{ji} = [(1/R_{ji+}) - (1/R_{ji-})]/(4\pi);$$

where R_{ji+} and R_{ji-} expressed in terms of distance $\bar{\ell}$ are the distances from the source and sink to P respectively; and m_{ji} is the strength of the jith dipole.

Similarly, since $\nabla(1/R) = -\mathbf{a}_R(1/R^2)$ where $\mathbf{a}_R = \mathbf{R}/R$, the magnetic flux density at P can be found from (3.6):

$$\mathbf{B} = \sum_{j=0}^{k}\sum_{i=0}^{n} m_{ji}\boldsymbol{\beta}_{ji} = \underline{\boldsymbol{\beta}}^T \underline{m} \tag{3.6}$$

where $\underline{\boldsymbol{\beta}}^T = [(\boldsymbol{\beta}_{00} \cdots \boldsymbol{\beta}_{0n})\ (\boldsymbol{\beta}_{10} \cdots \boldsymbol{\beta}_{1n})\ (\cdots)\ (\boldsymbol{\beta}_{k0} \cdots \boldsymbol{\beta}_{kn})];$ and $\boldsymbol{\beta}_{ji} = -\frac{\mu_o}{4\pi}\left(\frac{\mathbf{a}_{Rji+}}{R_{ji+}^2} - \frac{\mathbf{a}_{Rji-}}{R_{ji-}^2}\right).$

Note that (3.5) and (3.6) are in matrix form. For a DMP model that has a single dipole along the magnetization axis,

$$\varphi_{j0} = \varphi_{0i} = \begin{cases} 0 & i,j \neq 0 \\ \varphi_0 & i = j = 0 \end{cases} \text{ and } \beta_{j0} = \beta_{0i} = \begin{cases} 0 & i,j \neq 0 \\ \beta_0 & i = j = 0 \end{cases}$$

In (3.5) and (3.6), the unknown parameters (k, n, $\bar{\ell}$ and m_{ji}) characterize an appropriate distribution of dipoles which best approximate the field solutions.

To facilitate the design and control of PM-based devices, closed-form solutions are derived to describe the magnetic fields outside the physical region of the magnet, particularly near its boundary. The problem is to find an appropriate dipole distribution to best approximate the field solutions. The unknowns (k, n, $\bar{\ell}$ and m_{ji}) are solved minimizing the error function (3.7) subject to constraints imposed by the magnet geometry and a limited set of known field solutions (as fitting points):

$$E = \int_z [\Phi(z) - \Phi_A(z)]^2 dz \tag{3.7}$$

where $\Phi_A(z)$ is a known solution derived analytically, or curve-fit from solved numerical solutions or measured experimental data along the magnetization axis (say, the z-axis). The general expression [5] of the magnetic scalar potential Φ_A created at $\mathbf{R}'(x', y', z')$ to the field point $\mathbf{R}(x, y\ z)$ consists of a volume integral and a surface integral over the boundary surface S of the body volume V:

$$\Phi_A = \frac{1}{4\pi} \int_V \frac{-\nabla \bullet \mathbf{M}}{|\mathbf{R} - \mathbf{R}'|} dV + \frac{1}{4\pi} \int_S \frac{\mathbf{M} \bullet \mathbf{n}}{|\mathbf{R} - \mathbf{R}'|} dS \qquad (3.8)$$

where $\mathbf{n}$ is the unit surface normal. The corresponding magnetic flux density $\mathbf{B}$ can be found using (3.2).

The constraints are formed from a set of (known or pre-computed) solutions at specified points. For example, the residual magnetic flux density $B(z=z_o)$ of a PM (in manufacturer data sheet) can be specified as a constraint:

$$B(z_o) = B_A(z_o) = -\mu_o \nabla \Phi_A \big|_{z=z_o} \qquad (3.9)$$

(3.9) can be expressed in terms of the dipoles using (3.6). Since (3.7) accounts for the potential field along the magnetization axis, the $(k \times n + 1)$ constraints include (3.9) and the potential along two other orthogonal directions from (3.5):

$$\begin{bmatrix} \beta_0^T \big|_{z=z_o} \\ \varphi_1^T \\ \vdots \\ \varphi_{(k \times n)}^T \end{bmatrix} m = \begin{pmatrix} B_A(z_o) \\ \Phi_{A1} \\ \vdots \\ \Phi_{A(k \times n)} \end{pmatrix} \qquad (3.10)$$

In (3.10), the subscript "0" denotes that the dipole is along the magnetization vector. If the known fields are in terms of the measured magnetic flux density, Eq. (3.6) instead of (3.5) can be used to obtain (3.10). For PM-based actuator applications, the $\mathbf{B}$ and Φ values in (3.10) are evaluated at an appropriate magnet surface. To avoid the singularity at $\mathbf{R} = \mathbf{R}'$, the boundary can be approximated as

$$|\mathbf{R}| = \lim_{\varepsilon_R \to 0} \left(|\mathbf{R}'| \big|_{\text{point at surface}} + \varepsilon_R \right)$$

where ε_R is a small positive number. Three specific cases are worthy of mentioning:

i. If the relative permeability of the magnet is very large, the magnet surface can be approximated as equal potential, and Φ_A in (3.10) is a constant.
ii. If $\mathbf{M}$ is a constant implying $\nabla \bullet \mathbf{M} = 0$, the volume integral in (3.8) is zero and the potential field can be computed from the surface integral in (3.8).

iii. If the magnet is axi-symmetric, the magnetic field is uniform in a particular direction (say θ). To minimize the field variation in the θ direction when modeling with a finite number of dipoles, the following constraint can be imposed:

$$\left.\frac{Max[\Phi(\theta)] - Mean[\Phi(\theta)]}{Mean[\Phi(\theta)]}\right|_{\text{at the edge}} \times 100\% \leq \varepsilon_\theta \qquad (3.11)$$

where ε_θ is a specified (positive) error bound.

The DMP modeling method will be illustrated with examples in the following sections. In general, the unknown parameters (k, n, $\bar{\ell}$ and m_{ji}) of a DMP model can be determined following the procedure summarized below:

Step 1: Compute Φ_A and $\mathbf{B}_A$ analytically along the magnetization vector from (3.8) and (3.2) respectively.

Step 2: Generate an initial set of spatial grid points (k, n).

Step 3: Formulate (3.5) and (3.6) in terms of the unknowns, $\bar{\ell}$ and m_{ji}.

Step 4: Find $\bar{\ell}$ and m_{ji} by minimizing (3.7) subject to the constraint (3.10) where $\underline{\beta}_0^T$ and $\underline{\varphi}_I^T$ are obtained from (3.6) and (3.5) respectively. Error computed by (3.7) is saved.

Step 5: Check the condition (3.11). If (3.11) is not satisfied, increase k or n, and repeat from Step 3. Once (3.11) is met, the optimal parameters (k, n, $\bar{\ell}$ and m_{ji}) can be obtained by minimizing (3.7) using Step 4.

3.1.2 Numerical Illustrative Examples

As an illustration, two different types of PMs (cylindrical and wedge-shaped) are modeled using the DMP method for magnetic field computation.

Example 1: Cylindrical PMs
Cylindrical permanent magnets and electromagnets are commonly used, and some analytical solutions and/or experimental results are available for comparison. Thus, they are used here to illustrate the DMP modeling procedure. To facilitate practitioners in design, the formulation is expressed in dimensionless forms.

Consider a cylindrical magnet (radius a, length ℓ and $\mathbf{M} = M_o\mathbf{e}_z$) as shown in Fig. 3.2. The magnetic potential and flux density field solutions along the z-axis are given in [5]:

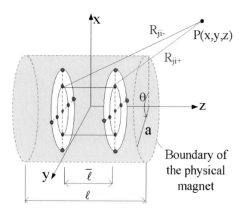

Fig. 3.2 DMP model of a cylindrical magnet

$$\frac{\Phi_A(Z)}{M_o \ell} = \frac{1}{4}[(A_- - |B_-|) - (A_+ - |B_+|)] \tag{3.12}$$

$$\frac{B_A(Z)}{\mu_o M_o} = \frac{1}{2}\left[\frac{|B_+|}{A_+} - \frac{|B_-|}{A_-} + c\right] \quad \text{where} \quad c = \begin{cases} 0 & \text{if } |Z| \geq 1 \\ 2 & \text{if } |Z| < 1 \end{cases} \tag{3.13}$$

where $Z = 2z/\ell$; $\gamma = 2a/\ell$; $A_\pm = \sqrt{\gamma^2 + B_\pm^2}$ and $|B_\pm| = |Z \pm 1|$.

The PM is modeled with k circular loops of n dipoles parallel to the magnetization vector as shown in Fig. 3.2. The loops (each with radius $\bar{a}_j$) are uniformly spaced:

$$\bar{a}_j = aj/(k+1) \quad \text{at } z = \pm \bar{\ell}/2 (0 \leq j \leq k) \tag{3.14}$$

$$0 < \bar{\ell} < \ell \tag{3.15}$$

For a cylindrical magnet, the field is uniform circumferentially and hence $m_{ji} = m_j$. To minimize the field variation in the θ direction, the constraint (3.11) evaluated at $r = a$ and $z = \ell/2$ is imposed. The unknowns ($\bar{\ell}$, m_j, n and k) are solved by minimizing (3.7) with Φ_A and Φ given by (3.8) and (3.5) respectively subject to the constraints imposed by (3.9)–(3.11), (3.14) and (3.15). For the DMP model shown in Fig. 3.2,

$$R_{ji\pm}^2 = [x - \bar{a}_j \cos i\theta_n]^2 + [y - \bar{a}_j \sin i\theta_n]^2 + (z \mp \bar{\ell}/2)^2 \tag{3.16a}$$

$$\frac{\mathbf{a}_{Rji\pm}}{R_{ji\pm}^2} = \frac{(x - \bar{a}_j \cos i\theta_n)\mathbf{a}_x + (y - \bar{a}_j \sin i\theta_n)\mathbf{a}_y + (z \mp \bar{\ell}/2)\mathbf{a}_z}{\left[(x - \bar{a}_j \cos i\theta_n)^2 + (y - \bar{a}_j \sin i\theta_n)^2 + (z \mp \bar{\ell}/2)^2\right]^{3/2}} \tag{3.16b}$$

where $i\theta_n$ indicates the angular position of the ith dipole on the jth loop and $\theta_n = 2\pi/n$.

3.1 Distributed Multi-pole Model for PMs

To provide some insight and for clarity in illustration, the following two cases are compared:

Case A: Single dipole model
The simplest approximation is to model the field with a single dipole at $x = y = 0$, which can be derived from (3.5),

$$\frac{\Phi(Z)}{(m/\ell)} = \frac{1}{2\pi}\left(\frac{1}{|Z-\delta|} - \frac{1}{|Z+\delta|}\right) \qquad (3.17a)$$

and

$$\frac{\mathbf{B}(Z)}{\mu_o m/\ell^2} = \frac{4\delta Z}{\pi(Z^2-\delta^2)^2}\mathbf{a}_z \qquad (3.17b)$$

where $\delta = \bar{\ell}/\ell$. However, as shown in (3.17a) and (3.17b) the single dipole model cannot account for the shape of the PM (or more specifically, the aspect ratio γ of the cylindrical PM).

Case B: DMP model (n = 4)
To account for the shape,

$$\frac{\Phi(Z)}{m_0/\ell} = \sum_{j=1}^{k}\frac{2m_j}{\pi m_0}\left(\frac{1}{\bar{A}_{j-}} - \frac{1}{\bar{A}_{j+}}\right) + \frac{1}{4\pi}\left(\frac{1}{|\bar{Z}_-|} - \frac{1}{|\bar{Z}_+|}\right) \qquad (3.18a)$$

$$\frac{\mathbf{B}(Z)}{\mu_o m_o/\ell^2} = \frac{1}{\pi}\sum_{j=1}^{k}\frac{m_j}{m_o}\left[\frac{\bar{Z}_-}{\bar{A}_{j-}^3} - \frac{\bar{Z}_+}{\bar{A}_{j+}^3}\right] + \left(\frac{1}{\bar{Z}_-^2 - \bar{Z}_+^2}\right) \qquad (3.18b)$$

where $\bar{A}_{j\mp} = \sqrt{\bar{\gamma}_j^2 + (\bar{Z}_\mp)^2}$; $\bar{Z}_\mp = Z \mp \delta$; and $\bar{\gamma}_j = 2\bar{a}_j/\ell$.

The results are given in Figs. 3.3, 3.4, 3.5, and Table 3.1 where

$$\%\text{Error} = 100 \times \int_z |\Phi(z) - \Phi_A(z)|dz / \int_z |\Phi_A(z)|dz.$$

Results in Table 3.1 were computed using MATLAB Optimization Toolbox. Since (3.8) is singular at the surface of a magnet, the Φ_A values for (3.10) are solved numerically with $|\mathbf{R}'| + 10^{-6}$; no significant difference in results was found when $\varepsilon_R \leq 10^{-3}$.

Some other observations are discussed as follows:

1. For a given aspect ratio $\gamma = 2a/\ell$, the parameters $\delta = \bar{\ell}/\ell$ and m_j/m_o can be calculated. The results for $\gamma \leq 1$ (with $k = 1$ and $n = 4$) are given in Fig. 3.4a, b. Once k, n, δ, and m_j/m_o are known, m_o can be determined for a specified B_r. In

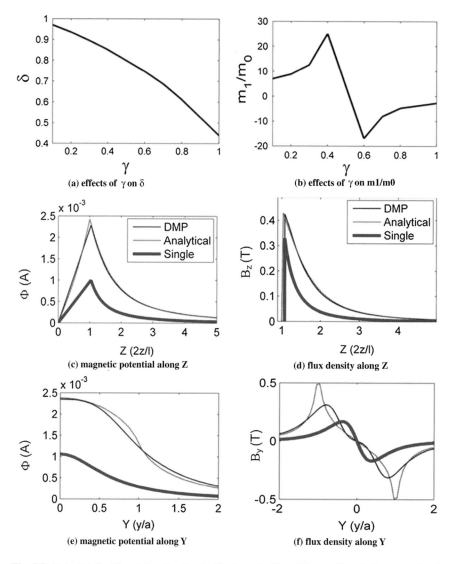

Fig. 3.3 Potential (in A) and flux density (in T) along the Y and Z axes (for all plots: $n = 4$ and $k = 1$; and for (**c–f**) $\gamma = 2a/\ell = 1$)

Fig. 3.3a, δ decreases as γ increases as expected. Given δ, the curve m_1/m_o in Fig. 3.3b depends only on the 1st term of (3.18a) and (3.18b). Figure 3.4b also shows that the case $m_1/m_o = 0$ (or only a single dipole) is very limited.

2. The DMP model is compared against the single dipole model and analytical solution in Fig. 3.3c–f. The analytical solution agrees well with the DMP model that uses only five dipoles ($n = 4$ and $k = 1$) to characterize the potential field and flux density of a PM with a unity aspect ratio ($\gamma = 1$). The single dipole, on the other hand, only provides a reasonable estimate of the magnetic flux density along the z-axis.

3.1 Distributed Multi-pole Model for PMs

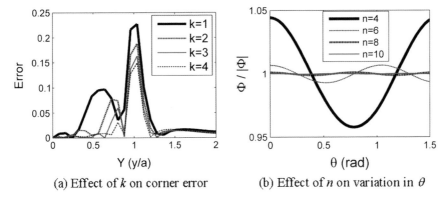

(a) Effect of k on corner error

(b) Effect of n on variation in θ

Fig. 3.4 Effect of n and k on modeling errors ($\gamma = 2a/\ell = 1$)

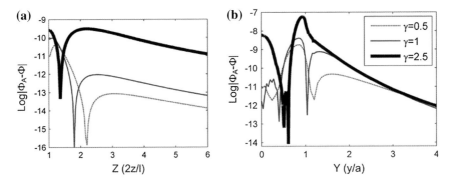

Fig. 3.5 Effect of the aspect ratio on modeling errors ($n = 4$; for $\gamma = 2a/\ell \leq 1$, $k = 1$; for $\gamma = 2.5$, $k = 2$)

Table 3.1 Parameters values

Model	$\delta = \bar{\ell}/\ell$	m (A m)	% error
Single ($n = k = 0$)	0.20	206.91	55.49
DMP ($n = 4$, $k = 1$)	0.39	$m_o = -33.31$; $m_{1i} = 78.13$	1.76
$\mu_0 M_0 = 1$ T, $\ell = 12.7$ mm, $\varepsilon_R = 10^{-6}$			

3. The discrepancy in Fig. 3.3f occurs primarily around the corner ($r = a$, $z = \ell/2$) of the PM; the errors in the magnetic flux density can be reduced by using more loops k. illustrates the effects of n and k on modeling accuracy.
4. Figure 3.4b shows the effect of increasing n in the circular loop, which effectively improves the uniformity circumferentially. The variation is about 5% with only five dipoles ($n = 4$ and $k = 1$) and nearly eliminated with $n \geq 6$.

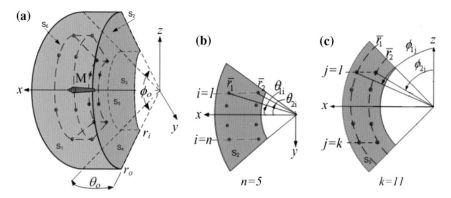

Fig. 3.6 Customized PM geometry (r_o = 46.5 mm, r_i = 23 mm, θ_o = 40°, ϕ_o = 70°; $\mu_o M_o$ = 0.62 T)

5. As shown in Fig. 3.5a, b where the absolute differences between the analytical and DMP modeled potentials are graphed in $\log_{10}$ scale, the model (with an increase in k) can be extended to account for the effect of larger aspect ratios.

Example 2: Customized wedge-shape PM

Figure 3.6 shows an example of customized PM used in a prototype spherical motor [6] where the magnet (shaped as a segment of a sphere) was magnetized in the positive x-axis. The potential field solution can be derived analytically from (3.8). Since $\mathbf{M} = M_o \mathbf{e}_x$ implying $\nabla \bullet \mathbf{M} = 0$, the volume integral in (3.8) is zero. With the differential surface and unit normal for each of the surfaces given in Table 3.2 Parameters for surface integral (3.8) where spherical coordinates (r, θ, ϕ) are defined in Fig. 3.1, the potential field can be computed from the surface integral in (3.8).

As shown in Fig. 3.6, the dipoles of the DMP model for the customized PM are uniformly located in a $\bar{r}_1 \times \bar{r}_2$ region where $r_o > \bar{r}_1 > \bar{r}_2 > r_i$ such that they form a $k \times n$ lattice as defined in Fig. 3.6b, c. Their locations are given by

Table 3.2 Parameters for surface integral (3.8)

Surfaces	Differential	Surface normal
$r = r_o$ $r = r_i$	$dS_{1,3} = r_{o,i}^2 \sin\phi \, d\theta \, d\phi$	$\hat{n}_1 = -\hat{n}_3 = \mathbf{e_r}$ $= \sin\phi \cos\theta \mathbf{e_x} + \sin\phi \sin\theta \mathbf{e_y} + \cos\phi \mathbf{e_z}$
$\phi = \phi_2, \phi = \phi_4$	$dS_{2,4} = r \, dr \, d\theta$	$\hat{n}_2 = -\sin\phi_1 \mathbf{e_x} + \cos\phi_1 \mathbf{e_z}$ $\hat{n}_4 = -\sin\phi_2 \mathbf{e_x} + \cos\phi_2 \mathbf{e_z}$
$\theta = \theta_5, \theta = \theta_6$	$dS_{5,6} = r \sin\phi \, dr \, d\phi$	$\hat{n}_5 = -\sin\theta_1 \mathbf{e_x} + \cos\theta_1 \mathbf{e_y}$ $\hat{n}_6 = \sin\theta_2 \mathbf{e_x} - \cos\theta_2 \mathbf{e_y}$

3.1 Distributed Multi-pole Model for PMs

Table 3.3 Parameters of the 11×5 DMP model (m_{ji}, where $j = 1, \ldots, 6$)

$i = 1$	7.0594	−0.0453	4.1249	2.0897	3.2027	2.5187
$i = 2$	−0.9223	−1.0204	−0.1404	−0.7011	−0.2487	−0.5497
$i = 3$	1.7825	0.3004	1.0298	0.6462	0.8940	0.7231

$\delta_1 = 0.3951$, $\delta_2 = 0.1702$

$$\phi_{2j} = j\phi_o/(k+1); \quad \theta_{2i} = i\theta_o/(n+1) \quad (3.19\text{a, b})$$

Since the dipoles are parallel to the *x*-axis, from Fig. 3.6c we have $\bar{r}_1 \cos \phi_{1j} = \bar{r}_2 \cos \phi_{2j}$ or

$$\phi_{1j} = \cos^{-1}(\bar{r}_2 \cos \phi_{2j}/\bar{r}_1) \quad (3.20\text{a})$$

Similarly, from the projections on the *x*-axis in Fig. 3.6b,

$$\theta_{1i} = \sin^{-1}\left[\frac{\bar{r}_2 \sin \phi_{2j}}{\bar{r}_1 \sin \phi_{1j}} \sin \theta_{2i}\right] \quad (3.20\text{b})$$

Since the magnet is symmetric about *xy* and *xz* planes, only one quarter of the dipole moments m_{ji} are found numerically using the optimization toolbox in MATLAB. The values are given in Table 3.3. To allow for one more degree of freedom (DOF) to describe the location of the dipoles, the source-sink spacing is defined using two variables:

$$\delta_1 = \frac{\bar{r}_1 - r_c}{r_o - r_c} \quad \text{and} \quad \delta_2 = \frac{\bar{r}_2 - r_c}{r_o - r_c}; \quad \text{where } r_c = \frac{r_o + r_i}{2} \quad (3.21)$$

The customized magnet was studied experimentally in [6, 7] giving measured magnetic flux density along the *x*-axis, and the ϕ direction in the *y-z* plane. These published data are used here as a basis for comparing the computed flux density of the customized magnet among three different models in Fig. 3.7.

(i) *Analytical integral* (3.8) with $\mathbf{M} = M_o \mathbf{e}_x$ and Table 3.2.
(ii) *Analytical integral (r-only)*: same as (i) but neglects ϕ and θ components of the magnetization; this model assumes a uniform radial field $\mathbf{M} \approx M_o \cos \theta \sin \phi \mathbf{e}_r$ [6].
(iii) 11×5 *DMP model* (Table 3.3) and $\mathbf{M} = M_o \mathbf{e}_x$ as in (i).

Some observations from the comparisons are summarized:

- As shown in Fig. 3.7a, the computed $B_x(x, 0, 0)$ of the DMP model along the *x*-axis closely agrees with both the analytical solution and experimental data. Note that since $\mathbf{M} = M_o \mathbf{e}_x$, $B_y = B_z = 0$ along the *x*-axis.

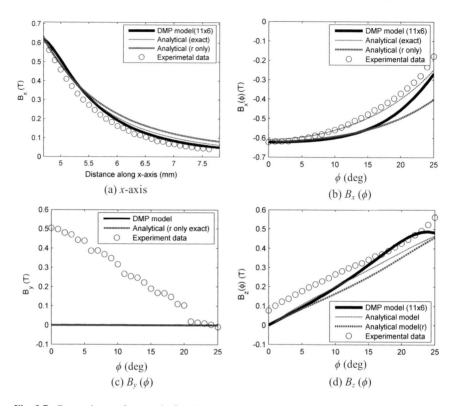

Fig. 3.7 Comparisons of magnetic flux density

- The three components of **B** that were measured at locations ($r = r_o + 0.5$ mm, $\theta = 0$) in terms of ϕ are given in Fig. 3.7b–d. The DMP computed B_x and the results of the first analytical integral agree well with the measured B_x data.
- $B_y(\phi)$ and $B_z(\phi)$ of the DMP model lie between the analytical integral model and the experimental data. Some discrepancies are observed between the computed and measured $B_y(\phi)$ and $B_z(\phi)$, which may be explained as follows. Since Φ is symmetric about the y-z plane, analytically we should have

$$B_y = -\mu_0 [\partial \Phi / \partial y]_{y=0} = 0 \quad \text{and} \quad B_z|_{y=z=0} = 0.$$

However, the measured B_y in Fig. 3.7c and B_z at $y = z = 0$ in Fig. 3.7d are not zero suggesting that the customized PM may not be uniformly magnetized or that there could be some systematic errors in measured B_y and B_z.

3.2 Distributed Multi-pole Model for EMs

The magnetic field of a multilayer (ML) EM can be characterized by the DMP model by treating it as a PM. The process of modeling an multi-layer EM as an ePM involves finding an equivalent magnetization **M** in terms of the current density **J** and geometry of the EM [8]. The magnetic flux density created at $\mathbf{R}'(x', y', z')$ to the field point $\mathbf{R}(x, y, z)$ is given by the Biot-Savart law:

$$\mathbf{B}_{EM} = \frac{\mu_0}{4\pi} \int_V \frac{\mathbf{J} \times (\mathbf{R} - \mathbf{R}')}{|\mathbf{R} - \mathbf{R}'|^3} dV \quad (3.22)$$

where μ_o is the free space permeability. For a PM, the magnetic flux density can be calculated from the negative gradient of the analytical magnetic potential [5]:

$$\mathbf{B}_{PM} = \frac{\mu_0}{4\pi} \int_V \frac{-(\nabla \bullet \mathbf{M})(\mathbf{R} - \mathbf{R}')}{|\mathbf{R} - \mathbf{R}'|^3} dV + \frac{\mu_0}{4\pi} \int_S \frac{(\mathbf{M} \bullet \mathbf{n})(\mathbf{R} - \mathbf{R}')}{|\mathbf{R} - \mathbf{R}'|^3} dS \quad (3.23)$$

where **n** is the unit surface normal. Unlike (3.22), the calculation of $\mathbf{B}_{PM}$ does not need the cross product of **J** and $\mathbf{R} - \mathbf{R}'$ vectors. Equations (3.22) and (3.23) provide the basis for deriving an ePM for the multilayer EM. The interest here is to seek the field solution outside the physical region of the EM, particularly near its boundary along the magnetization axis. The procedure is best illustrated through an example.

<u>Example: Cylindrical EM</u>
Cylindrical PMs and EMs are commonly used in motion systems. Some analytical and experimental results are also available for comparison. They are used here for clarity to illustrate the DMP modeling procedure. Figure 3.8a, b shows the

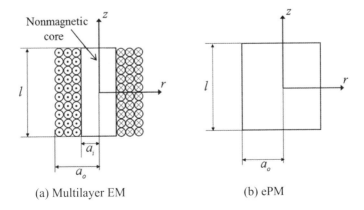

(a) Multilayer EM   (b) ePM

Fig. 3.8 Multilayer EM and equivalent PM

geometry of the cylindrical EM and its corresponding ePM (with the same l and a_o). The current density of the EM is given by (3.24):

$$\mathbf{J} = J(r)\mathbf{e}_\theta \quad \text{where} \quad \begin{cases} J(r) = 0, & 0 \leq r < a_i \\ J(r) = J, & a_i \leq r < a_o \end{cases} \quad (3.24)$$

where a_i and a_o are the inner and outer EM radii.

From (3.22), the z-component of the EM flux density can be calculated:

$$\frac{B_{EMz}(X,Y,Z)}{\mu_0 J(l/2)} = \frac{1}{4\pi} \int_{a_r}^{1} \int_{-1}^{1} \int_{0}^{2\pi} \frac{\rho(\rho - X\cos\theta - Y\sin\theta) d\theta dZ' d\rho}{\left[(X - \rho\cos\theta)^2 + (Y - \rho\sin\theta)^2 + L^2(Z - Z')^2\right]^{3/2}} \quad (3.25)$$

where $(X, Y, Z) = (x/a_o, y/a_o, 2z/l)$; $\rho = r/a_o$; $a_r = a_i/a_o$; and $L = l/(2a_o)$. A general closed-form solution to (3.25) is not available. To investigate the effect of the aspect ratios (a_r and L) on B_{EMz} for actuator design, (3.25) is numerically integrated at $z = l/2 + \varepsilon$ along the radial direction, where ε is a positive number. The results are graphed in Fig. 3.9. In Fig. 3.9a, the values are normalized to B_{EMz0}, or the value of B_{EMz} at $(0, 0, Z = 1 + 2\varepsilon/l)$, given in (3.26):

$$\frac{B_{EMz0}}{\mu_0 J a_o} = L \ln\left(\frac{1 + \rho_{o-}}{a_r + \rho_{i-}}\right) + L\left(\frac{\varepsilon}{l}\right) \ln\left[\frac{(1 + \rho_{o-})(a_r + \rho_{i+})}{(1 + \rho_{o+})(a_r + \rho_{i-})}\right] \quad (3.26)$$

where $\rho_{o+} = \sqrt{4L^2(\varepsilon/l)^2 + 1}$; $\rho_{o-} = \sqrt{4L^2(1 + \varepsilon/l)^2 + 1}$; $\rho_{i+} = \sqrt{4L^2(\varepsilon/l)^2 + a_r^2}$; and $\rho_{i-} = \sqrt{4L^2(1 + \varepsilon/l)^2 + a_r^2}$. When $\varepsilon/l \ll 1$ or near the physical boundary,

$$\left.\frac{B_{EMz0}}{\mu_0 J l/2}\right|_{(\varepsilon/l) \to 0} = \ln\left(\frac{1 + \sqrt{1 + 4L^2}}{a_r + \sqrt{a_r^2 + 4L^2}}\right) \quad (3.27)$$

Some observations can be made in Fig. 3.9:

- As shown in Fig. 3.9a, b, B_{EMz} linearly decreases from a_i to a_o along the radial direction. When $0.25 \leq L \leq 1$, B_{EMz} is relatively uniform inside the air core. B_{EMz0} increases with coil thickness (or smaller a_r) for the same a_o and l implying that thicker coils have higher magnetic fluxes (proportional to the area under the curve).
- Figure 3.9c shows that the drop in B_{EMz0} is approximately linear with a_r. B_{EMz0}, however, increases exponentially with L and approaches a constant for a given a_r, Fig. 3.9d.

3.2 Distributed Multi-pole Model for EMs

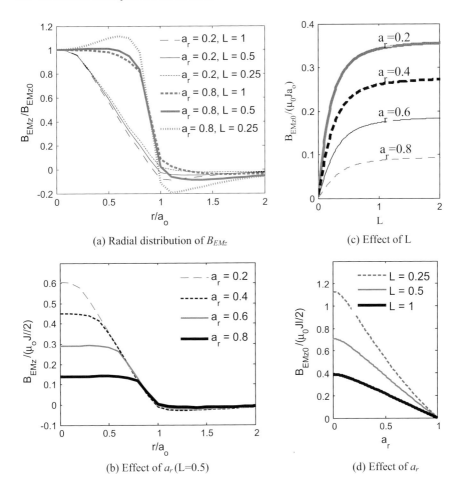

Fig. 3.9 Effect of a_r and L and on B_{EMz0} ($\varepsilon/l = 0.01$)

3.2.1 Equivalent Magnetization of the ePM

For a cylindrical PM, **M** is zero outside the physical boundary where $r \geq a_o$. This and the above observations suggest that the magnetization of the ePM takes the form

$$\mathbf{M} = M(r)\mathbf{e_z} \quad \text{where} \quad \begin{cases} M(r) = M_o, & 0 \leq r < a_i \\ M(r) = M_o - J(r - a_i), & a_i \leq r \leq a_o \end{cases} \quad (3.28)$$

where M_0 is an integral constant to be found by comparing (3.22) and (3.23). Since the cylindrical ePM has a maximum along its magnetization, M_0 can be found from

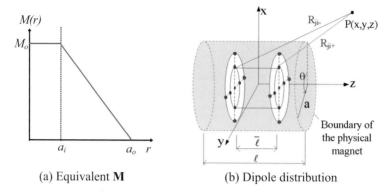

Fig. 3.10 DMP model of a magnet

$B_{PMz} = B_{EMz}$ at $(0, 0, l/2 + \varepsilon)$. Substituting (3.28) into (3.22) and noting that $\nabla \cdot \mathbf{M} = 0$, the first term on the right side of (3.24) disappears, and the second term can be written as:

$$\frac{B_{PMz0}}{\mu_0 Jl/2} = \frac{B_{EMz0}}{\mu_0 Jl/2} + \frac{1}{Jl}[J(a_o - a_i) - M_0]\left(\frac{\varepsilon}{\rho_{o+}a_o} - \frac{l+\varepsilon}{\rho_{o-}a_o}\right) \quad (3.29)$$

M_0 can now be determined by equating the last term of (3.29) to zero such that $B_{PMz0} = B_{EMz0}$. As the factor involving the independent variable ε is not always zero, $M(r) = J(a_o - a_i)$. Hence, the equivalent magnetization $\mathbf{M}$ graphically illustrated in Fig. 3.10a is given by (3.30):

$$\mathbf{M} = M(r)\mathbf{e_z} \quad \text{where} \quad \begin{cases} M(r) = J \cdot (a_o - a_i), & 0 \leq r < a_i \\ M(r) = J \cdot (a_o - r), & a_i \leq r \leq a_o \end{cases} \quad (3.30)$$

Since J is uniform throughout the entire volume of an EM, substituting (3.30) into (3.23) yields:

$$\mathbf{B}_{ePM} = \frac{J\mu_0}{4\pi} \int_V \frac{-(\nabla \cdot \hat{\mathbf{M}})(\mathbf{R} - \mathbf{R}')}{|\mathbf{R} - \mathbf{R}'|^3} dV + \frac{\mu_0}{4\pi} \int_S \frac{(\hat{\mathbf{M}} \cdot \mathbf{n})(\mathbf{R} - \mathbf{R}')}{|\mathbf{R} - \mathbf{R}'|^3} dS \quad (3.31a)$$

where

$$\hat{\mathbf{M}} = \mathbf{M}/J \quad (3.31b)$$

Once the ePM is found with the equivalent magnetization (3.31b), the EM can be modeled using a distributed set of multi-poles (DMP). For a cylindrical EM, the DMP consists of k circular loops of n equally spaced dipoles parallel to the magnetization vector as shown in Fig. 3.3b. The loops (each with radius $\bar{a}_j$) are uniformly spaced:

3.2 Distributed Multi-pole Model for EMs

$$\bar{a}_j = a_o j/(k+1) \quad \text{at} \quad z = \pm \bar{\ell}/2, \quad \text{where} \quad 0 \leq j \leq k \tag{3.32}$$

The flux density at point P(x, y, z) generated by an EM can be computed using:

$$\mathbf{B}_{EM} = \frac{\mu_0}{4\pi} \sum_{i=0}^{k} m_i \sum_{j=1}^{n} \left(\frac{\mathbf{R}_{ij+}}{|\mathbf{R}_{ij+}|^3} - \frac{\mathbf{R}_{ij-}}{|\mathbf{R}_{ij-}|^3} \right) \tag{3.33}$$

where $\mathbf{R}_{ij+}$ and $\mathbf{R}_{ij-}$ are the vectors from the source and sink of the jth dipole on the ith loop to **P** respectively; the procedure of deriving the parameters (k, n, $\bar{l}$ and m_i) with the equivalent magnetization (3.30) follow the same steps for a PM as illustrated in Sect. 3.1.1.

3.2.2 Illustrations of Magnetic Field Computation

The flux densities of an multi-layer EM computed using DMP_{EM} model are compared with those computed with two commonly used methods:

(1) Numerically integrate the exact integral (3.25) for the flux density of a multi-layer (ML) EM;
(2) Equivalent single layer (ESL) model [1, 9] with approximated magnetic fields for ML EMs which simplifies the Lorentz force equation by reducing the volume integral to a surface integral.

Since the ESL model is singular at the surface, B_z and B_r are plotted along the radial direction at $z = l/2 + \varepsilon$ with $\varepsilon = 0.5$ mm, and B_z along the z-axis for the thick EM in Fig. 3.11. Table 3.4 lists the dimensions of the EM and the values of the parameters defining the ESL and DMP_{EM} models.

It can be seen that unlike the ESL model where the equivalent current density J_e is determined from the 2D magnetic field, the equivalent magnetization **M** of the ePM is derived using the complete 3D integral. As shown in Fig. 3.11, the DMP_{EM} modeled flux densities agree very well with the solutions to the exact integral for both thin and thick coils. The ESL model provides a reasonable prediction of the z-component flux density, but there are relatively large discrepancies from the exact solutions as the EM has a thick layer of winding (smaller a_i/a_o).

3.3 Dipole Force/Torque Model

3.3.1 Force and Torque on a Magnetic Dipole

Two methods commonly used in calculating the forces between stator EMs and rotor PMs of a spherical motor are the Lorentz force equation and the Maxwell

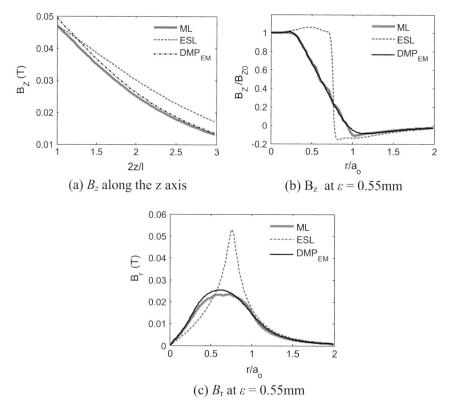

Fig. 3.11 B_{EMZ} in tesla

Table 3.4 Simulation parameters

1000 turns, #28 wire, 1 A current
ML: $a_o = 15.88$ mm, $a_r = 0.3$, $L = 0.3$
ESL: $J_e d_w = 120.1$ A/mm, $a_e = 12$ mm
DMP$_{EM}$: $n = 16$, $k = 6$, $\bar{l}/l = 0.442$
m_i (μA/m): 1.476, 0.547, 1.618, 1.644, 1.654, 1.325, 0.592

stress tensor [5]. These methods require solving the magnetic field and computing a volume or surface integral to derive the force model. As general closed-form solutions are not available, the volume or surface integrals are often solved numerically.

An alternative method to compute the magnetic force is the Lorentz force law in analogy to that on an electric charge as illustrated in Fig. 3.12, where a dipole (with strength m) is defined as a pair of source and sink separated by a finite distance. The force **F** and torque **T** acting on the dipole can be written (in analogy to that on a stationary electric charge by the Lorentz law) [10] as

3.3 Dipole Force/Torque Model

Fig. 3.12 Force on dipoles

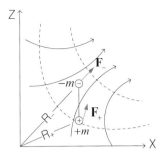

$$\mathbf{F} = \mu_o m[\mathbf{H}_{R+} - \mathbf{H}_{R-}] \quad (3.34a)$$

$$\mathbf{T} = \mu_o m[\mathbf{R}_+ \times \mathbf{H}_{R+} - \mathbf{R}_- \times \mathbf{H}_{R-}] \quad (3.34b)$$

where μ_0 is free space permeability; H_{R+} and H_{R-} are the magnetic field intensities acting on the magnetic source and sink of the dipole respectively, and the subscripts R_+ and R_- are the corresponding distances from a field point.

With both the PMs and EMs are modeled using the DMP method, the magnetic forces on the system can be calculated using the Maxwell stress tensor method or the dipole force equation. Unlike the commonly used Lorentz force equation and the Maxwell stress tensor method, the dipole force equation (replacing integrations with summations) dramatically reduces computation time. As will be shown, the closed form dipole model is an efficient way to compute the inverse torque model of an over-actuated system, especially for PMSMs where a large number of stator EMs and PMs are involved.

3.3.2 Illustration of Magnetic Force Computation

The computation process using DMP model is illustrated where the results and computational efficiencies are compared against known solutions. The results are given in Sections *A* and *B* followed by discussions in Section *C*.

The magnetic force between a PM and an EM for two test setups shown in Fig. 3.13 is computed. Published experimental force data [11] are available for comparison. In the following computation, the PMs are modeled as DMP_{PM} with the parameters summarized in Table 3.5.

Three methods for modeling the magnetic fields and forces are compared:

Method 1 computes the force using Maxwell Stress Tensor

$$\mathbf{F} = \oint_C \mathbf{T} dC \quad \text{where} \quad \mathbf{\Gamma} = \frac{1}{\mu_0}\left(\mathbf{B}(\mathbf{B} \bullet \mathbf{n}) - \frac{1}{2}B^2 \mathbf{n}\right) \quad (3.35)$$

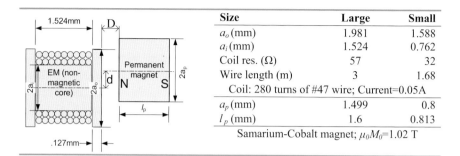

Fig. 3.13 Experimental setup [11] and parameters

Table 3.5 Simulation parameters

	Parameters	Large	Small
PM DMP$_{PM}$	**n, k,** $\bar{l}/l$	6, 2, 0.314	6, 2, 0.3122
	m_i (µA/m)	1.65, 0.02, 3.8	0.43, 0.02, 1.07
EM (ESL)	$J_e d_w$ (µA/mm)	22.75	38.98
	a_e (mm)	1.8168	1.456
EM (DMP$_{EM}$)	n, k, $\bar{l}/l$	12, 8, 0.7661	8, 3, 0.7441
	m_i (nA/m)	0.236, 0.177, 0.366, 0.567, 0.751, 0.914, 1.032, 1.28, 0.312	1.354, 1.758, 3.32, 1.661

where C is an arbitrary boundary enclosing the body of interest; and **n** is the normal of the boundary interface. (3.35) requires the total field **B** (contributed by both the PM and EM) to compute the force by the surface integration. As a basis for comparison, the **B**-field of the multilayer EM numerically computed using (3.22).

Method II calculates the Lorentz force exerted on the current carrying EM

$$\mathbf{F} = -\oint \mathbf{B} \times Id\mathbf{n} \quad \text{where } I = \oiint Jd S \quad (3.36)$$

where **n** is the unit current direction vector; and S is the cross section of wire. Since the current density vector **J** is directly used in the calculation, only the **B**-field of the PM is needed in the Lorenz force Eq. (3.36). The multilayer EM is replaced with the equivalent single layer (ESL) model (with equivalent current density J_e, wire diameter d_w, and coil radius a_e), which reduces the volume integral to a surface integral.

Method III uses the dipole force equation in analogy to that on a stationary electric charge by the Lorentz law to compute the net force acting on the PM.

The net force is simply the summation of the individual forces on the dipoles that characterize the PM:

3.3 Dipole Force/Torque Model

$$\mathbf{F} = \frac{J\mu_0}{4\pi} \sum_{i=1}^{n_r} m_{r_i} \sum_{j=1}^{n_s} m_{s_j} \left(\mathbf{R}_{s_{j+} r_{i+}} - \mathbf{R}_{s_{j+} r_{i-}} + \mathbf{R}_{s_{j-} r_{i-}} - \mathbf{R}_{s_{j-} r_{i+}} \right) \quad (3.37)$$

where $\mathbf{R}_{r_{i\pm} s_{j\pm}} = (\mathbf{R}_{r_{i\pm}} - \mathbf{R}_{s_{j\pm}})/|\mathbf{R}_{r_{i\pm}} - \mathbf{R}_{s_{j\pm}}|^3$ where $\mathbf{R}_{r_{i\pm}}(\mathbf{R}_{s_{j\pm}})$ is the *i*th (*j*th) pole location of the rotor (EM$_j$); the signs, (+) and (−), stand for the source and the sink of the dipole; n_r and n_s are the number of dipoles of the PM and EM; and m_{ri} (m_{sj}) are the pole strength of the *i*th (*j*th) dipole pair in the rotor (EM$_j$). The EM is modeled as DMP$_{EM}$.

The parameters for the ESL model and the DMP$_{EM}$ are summarized in Table 3.5. The magnetic fields of the large and small coils are given in the left and right columns in Fig. 3.14, where B_z is plotted along the z axis; and B_z and B_r are plotted along the radial direction at $z = l/2 + \varepsilon$ with $\varepsilon = 0.5$ mm. The computed forces F are compared against published experimental data F_{exp} in Fig. 3.15. Compares the time required to compute 26 data points in Fig. 3.15a using a computer with Quad Core 2.66 GHz CPU and 8 GB RAM.

Some observations on Figs. 3.14 and 3.15 are discussed as follows:

- Unlike the ESL model where the equivalent current density J_e is determined from the 2D magnetic field, the equivalent magnetization **M** of the ePM is derived using the complete 3D integral. As shown in Figs. 3.14 and 3.15, the DMP$_{EM}$ modeled flux densities agree very well with the solutions to the exact integral (3.25) for both thin and thick coils. The ESL model provides a reasonable prediction of the z-component flux density, but discrepancies from the exact solutions increase with coil thickness (or smaller a_i/a_o).
- The Maxwell stress tensor in Method I can be computed using the DMP$_{PM}$ and DMP$_{EM}$, which yields the same solution to the dipole force equation in Method III. However, unlike the Maxwell stress tensor method or the Lorentz force equation (with the ESL approximation) that require numerical computations of a surface integration, the dipole force equation (replacing integrations with summations) is in closed-form dramatically reducing computation to 0.0625 s as compared in Table 3.6.
- As shown in Fig. 3.15, the Maxwell stress tensor and the dipole force equation (or Methods I and III respectively) agree very closely with published experimental data while the ESL model (that reduces the volume integral of the multi-layer EM to a surface integral of a single-layer coil) overestimates the computed forces as expected.

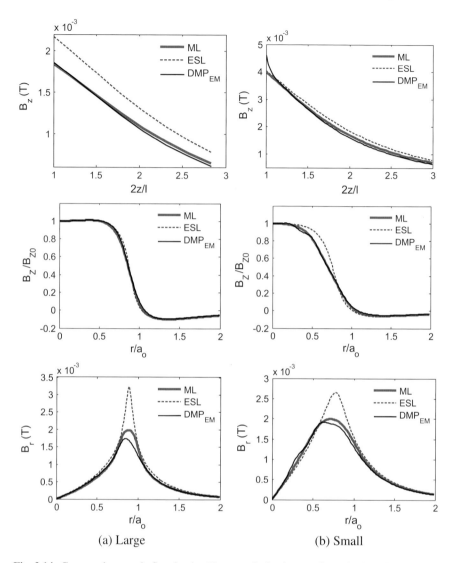

Fig. 3.14 Computed magnetic flux density (the numerical values as shown in the left column are provide in the appendix for reference)

3.4 Image Method with DMP Models

Existing techniques for analyzing electromagnetic fields when magnetically conducting boundaries are involved have difficulties in achieving both accuracy and low computation time simultaneously. The DMP method can be extended to overcome these difficulties to derive closed-form field solutions for design and motion control of the actuators. This section introduces the image method to

3.4 Image Method with DMP Models

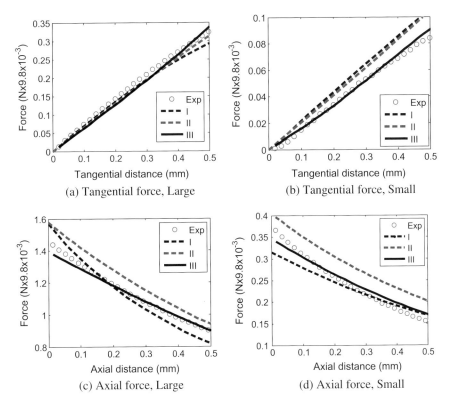

Fig. 3.15 Computed forces and experimental data (the numerical values as shown in the left column are provide in the appendix for reference)

Table 3.6 Comparison of computational times

Method	I	II	III
Computation time (s)	106.03	21.53	0.0625

account for the effects of magnetic conducting boundaries, which reduces the problems to a more tractable form for calculating the magnetic fields and torques [12, 13].

3.4.1 Image Method with Spherical Grounded Boundary

Consider here a class of electromagnetic problems where magnetic charges are in the presence of a magnetically grounded conducting boundary. Except at the point charges, the magnetic field is continuous and irrotational, for which a scalar potential Φ can then defined such that

$$\mathbf{H} = -\nabla\Phi; \quad \mathbf{B} = \mu_0 \mathbf{H}; \qquad (3.38\text{a, b})$$

where μ_0 is the permeability of free space. The formal approach for solving the magnetic field at every point outside the conducting boundary would be to solve the Laplace equation $\nabla^2 \Phi = 0$, the solution to which must satisfy the following conditions. At points very close to the magnetic charge (source or sink), the potential Φ approaches that of the point charge alone

$$\Phi \to \pm m/(4\pi R), \quad \text{as } R \to 0 \qquad (3.39)$$

where m is the strength of the magnetic charge; "+" and "−" designate that the pole is a source or a sink respectively; and R is the distance to $\pm m$. In addition, the potential is zero ($\Phi \to 0$) at the grounded conducting surface and points very far from $\pm m$.

Image Method of a Magnetic Charge

An alternative approach in lieu of a formal solution is the method of images, which replaces bounding surfaces by appropriate image charges, and the potential distributions can then be determined in a straightforward manner. As an illustration, consider Fig. 3.16a where the magnetic charge m (source or sink) in the free space is enclosed by the conducting spherical boundary (radius R) of very high permeability ($\mu \to \infty$, such as iron). The interest here is to determine the Φ distribution inside the grounded spherical surface due to the charge m.

Without loss of generality, the surface is assigned a constant zero potential or $\Phi = 0$. In Fig. 3.16a where XYZ is the reference coordinate system, $\bar{m}$ is the image of the charge m and lies along the radial line connecting m. The image charge must be outside the region in which the field is to be determined, the parameters (m, a) and $(\bar{m}, \bar{a})$ are related by (3.40):

$$\bar{a}/R = -\bar{m}/m = \Lambda \quad \text{where} \quad \Lambda = R/a. \qquad (3.40)$$

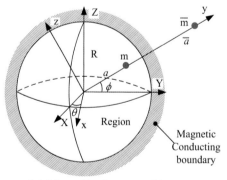
(a) Magnetic charge and image

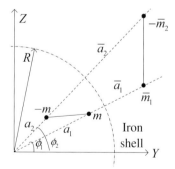
(b) Image of a dipole

Fig. 3.16 Spherical magnetic boundary

3.4 Image Method with DMP Models

To facilitate the discussion, a local coordinate frame xyz is defined such that m and $\bar{m}$ are on the y axis at the vector positions, $\mathbf{a}$ and $\bar{\mathbf{a}}$, respectively. In terms of spherical coordinates (r, θ, ϕ), any point $\mathbf{x}(x, y, z)$ in the local frame can be expressed in the reference XYZ frame:

$$\mathbf{X}/|\mathbf{x}| = [\cos\theta \cos\phi \quad \sin\theta \cos\phi \quad \sin\phi]^T \tag{3.41}$$

where $\theta = \tan^{-1}(y/|\mathbf{x}|)$; and $\phi = \cos^{-1}(z/|\mathbf{x}|)$. Due to the symmetry of a sphere, the problem can be reduced to two dimensional (2D) in the yz plane. The potential at point $\mathbf{p}(x, y, z)$ for $0 \leq p \leq R$ where $p = |\mathbf{p}|$ is given by

$$\Phi(\mathbf{p}) = \frac{m}{4\pi}\left(\frac{1}{\sqrt{p^2 + a^2 - 2\mathbf{p}\cdot\mathbf{a}}} - \frac{1}{\sqrt{(p/\Lambda)^2 + R^2 - 2\mathbf{p}\cdot\bar{\mathbf{a}}}}\right) \tag{3.42}$$

$a = |\mathbf{a}|$; and $\bar{a} = |\bar{\mathbf{a}}|$. It can be seen from (3.42) that when $p = R$ (on spherical surface), Φ vanishes. The solution is exactly the same as that between two point charges (m and $\bar{m}$) without the grounded spherical boundary.

Images of a Magnetic Dipole

Since magnetic poles exist in pairs, a *dipole* is defined as a pair of source m and sink $-m$ separated by a distance $\mathbf{d}$. For a dipole, the images of its source and sink (located at $\mathbf{a_1}$ and $\mathbf{a_2}$) are denoted as $\bar{m}_1$ and $-\bar{m}_2$ in Fig. 3.16b. Using (3.40), the potential at $\mathbf{p}$ in the free space containing the dipole can be expressed as

$$\Phi(\mathbf{p}) = \frac{m}{4\pi}\left[\left(\frac{1}{|\mathbf{r}_1|} - \frac{\Lambda_1}{|\bar{\mathbf{r}}_1|}\right) - \left(\frac{1}{|\mathbf{r}_2|} - \frac{\Lambda_2}{|\bar{\mathbf{r}}_2|}\right)\right] \tag{3.43}$$

where

$$\mathbf{r}_i = \mathbf{p} - \mathbf{a_i}; \quad \bar{\mathbf{r}}_i = \mathbf{p} - \bar{\mathbf{a}}_i \tag{3.44a, b}$$

$$\mathbf{a_i}/|\mathbf{a_i}| = \bar{\mathbf{a}}_i/|\bar{\mathbf{a}}_i| = [0 \quad \sin\phi_i \quad \cos\phi_i]^T \tag{3.45}$$

and $i = 1, 2$ denote the source and sink respectively. In general, if $\mathbf{a_1} \neq \mathbf{a_2}$, $\bar{m}_1 \neq \bar{m}_2$; as a result, the image source and sink of a dipole do not form a dipole, and do not satisfy the condition for continuous flow, $\nabla \cdot B = 0$. The solution of the image method is not valid in the magnetic conducting sphere since the image dipole does not actually exist but the images are rather standing in for the magnetic densities induced on the magnetic boundary.

Image Method for Spherical Actuators

For practical applications, the following cases are illustrated in spherical coordinates (r, θ, ϕ):

Case 1: The dipole m is outside the magnetically grounded spherical rotor of radius r_r, and its source and sink are located at $\mathbf{x}_{r1}$ and $\mathbf{x}_{r2}$ respectively:

$$|\bar{\mathbf{x}}_{ri}|/r_r = -\bar{m}_{ri}/m = \Lambda_{ri} \quad \text{where} \quad \Lambda_{ri} = r_r/|\mathbf{x}_{ri}| \qquad (3.46)$$

and $\mathbf{x}_{ri}/|\mathbf{x}_{ri}| = \bar{\mathbf{x}}_{ri}/|\bar{\mathbf{x}}_{ri}| = [\cos\theta_i \cos\phi_i \quad \sin\theta_i \cos\phi_i \quad \sin\phi_i]^T$.

Case 2: The source and sink of the dipole m are inside the hollow magnetically grounded spherical stator of radius r_s, and located at $\mathbf{x}_{s1}$ and $\mathbf{x}_{s2}$ respectively.

$$|\bar{\mathbf{x}}_{sj}|/r_s = -\bar{m}_{sj}/m = \Lambda_{sj} \quad \text{where} \quad \Lambda_{sj} = r_s/|\mathbf{x}_{sj}| \qquad (3.47)$$

and $\mathbf{x}_{sj}/|\mathbf{x}_{sj}| = \bar{\mathbf{x}}_{sj}/|\bar{\mathbf{x}}_{sj}| = [\cos\theta_j \cos\phi_j \quad \sin\theta_j \cos\phi_j \quad \sin\phi_j]^T$.

Case 3: The dipole m is in-between the grounded spherical rotor and stator, which are concentric. Each of the source and the sink will have two images located such that $\theta_i = \theta_j = \theta$ and $\phi_i = \phi_j = \phi$:

$$\bar{\mathbf{x}}_{ri}/|\bar{\mathbf{x}}_{ri}| = \bar{\mathbf{x}}_{sj}/|\bar{\mathbf{x}}_{sj}| = [\cos\theta \cos\phi \quad \sin\theta \cos\phi \quad \sin\phi]^T \qquad (3.48)$$

This is essentially a combination of Cases 1 and 2. Since the Laplace equation is linear, the solution can be obtained using the principle of superposition of Cases 1 and 2.

With the specified magnetic dipoles and boundary, the images of the sources and sinks can be calculated from (3.46) to (3.48), Φ and hence **H** in the free space can be found from (3.44) and (3.38a,b) respectively.

3.4.2 Illustrative Examples

Consider the electromagnetic system (Fig. 3.17) which consists of two axially magnetized PMs on the spherical rotor, and an air-cored EM on the inside surface of the hollow spherical stator. The two PMs are identical but their magnetization vectors are in opposite directions. The values of the parameters used in this simulation are given in Table 3.7. The interest here is to investigate the effects of the iron boundaries on the magnetic field distribution (in the region between the rotor and stator surfaces) and the torque acting on the rotor. To derive closed-form field solutions to facilitate the design and control, the field solutions are derived outside the physical region of the magnet and magnetic boundary, particularly in the free space where the EM is located. Using the DMP model, k circular loops of n dipoles (strength m_k) evenly spaced on the circular loop of radius ρ and parallel to the magnetization vector are used to model the PM. The parameters characterizing the DMP model are summarized in. The corresponding images (location and strength) reflecting the source and sink of each dipole on the spherical boundaries can be derived from (3.43) with (3.46) to (3.48).

3.4 Image Method with DMP Models

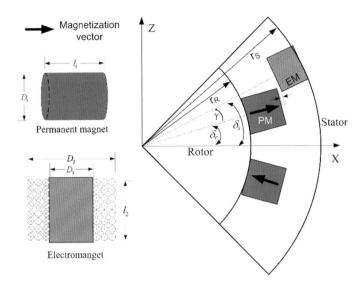

Fig. 3.17 Actuator system with two PMs and EM

Table 3.7 System parameters

Spheres	PM and its DMP model					EM
$r_r = 25.4$	$\ell_1 = 12.7$; $D_1 = 12.7$; $M_o = 1.34$ T					$\ell_2 = 25.4$
$r_s = 63.5$	k	n	d/l	$2\rho/D_1$	m_k ($\times 10^{-5}$ A m)	$D_2 = 19.05$
$\delta_s = 26°$	0	1	0.264	0	−2.29	$D_3 = 9.525$
$\delta_r = 20°$	1	6	0.514	0.5	6.18	$N = 1040$, $I = 4$ A

Dimensions in mm; N = # of turns

Figure 3.18 shows the magnetic field of the PM pair between two concentric magnetically grounded spheres when there is no current flowing through the air-core EM. To visually illustrate the image method, the effects of image dipoles in the grounded spheres is graphed. It must be emphasized that the field distributions calculated using the image method are valid only in the free space between the spheres, and are invalid in the region $a < r_r$ and $a > r_s$ where Φ and $\mathbf{H}$ are zero in iron ($\mu \to \infty$) and are veiled in Fig. 3.18.

The effects of iron boundaries on the magnetic field distribution due to the pair of PMs are investigated by comparing four different design configurations (DCs):

DC1: Rotor and stator are non-magnetic boundaries;
DC2: Only the rotor is a magnetically conducting sphere.
DC3: Only the stator is a magnetic conducting boundary.
DC4: Both the rotor and stator are magnetic boundaries.

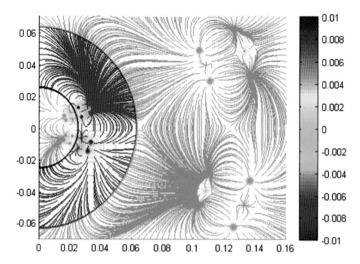

Fig. 3.18 PM pair between two magnetic surfaces

The simulated magnetic fields are compared in Fig. 3.19, where the bold solid circles indicate the spherical boundaries (black for the rotor and red for the stator). Figure 3.19a, where there are no magnetic boundaries, serves as a basis for comparison. The effects of the iron rotor and stator boundaries on the magnetic field are compared in Fig. 3.19b, c respectively. As expected, the magnetic field is perpendicular to the magnetically grounded spherical surface ($\Phi = 0$). Similarly, the combined effect of both the iron rotor and stator boundaries on the magnetic field I graphically displayed in Fig. 3.19d.

3.4.3 Effects of Iron Boundary on the Torque

Once the magnetic field is known, the torque acting on the rotor can be computed using the Lorentz force equation or Maxwell stress tensor.

Numerical Investigation
To validate the torque computed using the magnetic field computed from the DMP/image method, the torque of the electromagnetic system shown in Fig. 3.20a is computed, where the values of the parameters are based on, and compare the results against the numerical method using ANSYS.

In Fig. 3.20a, the rotor consisting of a pair of PMs rotates with respect to two stator-EMs. Both the diagonally symmetric PMs and EMs are mounted on spherical magnetic boundaries in the same plane. In this study, the torque about the axis perpendicular to this plane is computed as a function of the separation angle γ.

FE Model: Because of the symmetry, the finite element (FE) model in ANSYS uses cylindrical iron boundaries for simplicity. The procedure for computing the

3.4 Image Method with DMP Models

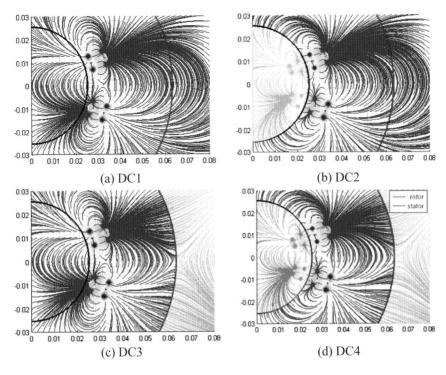

Fig. 3.19 Effect of boundaries on magnetic field

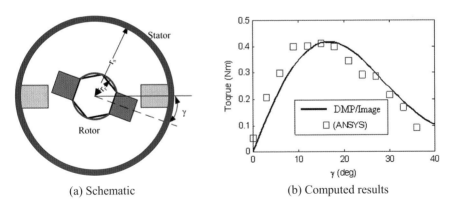

Fig. 3.20 Comparisons of computed torques (Computational time: ANSYS = 1480 s. DMP = 17 s)

electromagnetic torque using ANSYS can be found in [14, 15]. In ANSYS, the iron boundary was modeled using the eight-node SOLID96 elements ($\mu_r = 1000$ where μ_r is the relative permeability); the free space air volume was modeled using four-node INFIN47 elements; and the air-cored stator coil as SOURCE36 elements. For the PM, $\mu_o \mu_r = B_r/H_c$, where B_r and H_c are the residual magnetization and the magnitude of coercive force vector respectively. In this computation, $H_c = 795{,}770$ A/m and $B_r = 1.34$ T. With the total magnetic flux density obtained from ANSYS, the torque acting on the rotor is computed from Maxwell stress tensor (2.28) where Γ is a circular boundary enclosing the rotor with the pair of PMs.

Image Method with DMP Model: The two concentric spherical boundaries along with the DMP model of the PM pair are employed in finding the images of the sources and sinks of the dipoles as discussed in above sections. With the specified multi-dipoles and the corresponding images, the magnetic field in the air space between the conducting surfaces is computed, which serves as a basis for computing the Lorentz force and hence the torque from (3.34a) to (3.34b).

The computed torque is compared in Fig. 3.20b, which shows good agreement with a maximum error less than 5%. Using a Windows-based PC (dual-core processor 2.21 GHz CPU and 1 GB RAM), the image method with DMP model requires only 13 s to compute the torque curve while ANSYS requires 24.67 min to compute the thirteen data points. Some discrepancies occurred from the initial position. In the FE method, the quality of mesh significantly affects the accuracy of FE. The distortion of the automatically generated FE mesh could be the cause of the FE error (offset) even when the separation angle is zero.

Effects of Pole Design with Iron Boundary

The effects of iron boundaries on the actuator design are further examined by comparing the magnetic torque of the electromagnetic systems shown in for a given stator radius r_s. Table 3.8 summarizes the specific geometries of the two designs respectively, where the differences are highlighted in **bold**; and g is the radial air-gap between the air-cored EM and rotor PMs. The values of the parameters that characterize the DMP models of the PM and EM are given in Tables 3.9 and 3.10 respectively, where the % error is DMP modeling error defined along the magnetization axis of the PM:

Table 3.8 Comparison of design parameters

Spheres	Design A	Design B
r_r, r_s, g	**25.4**, 64.3, 0.76	**38.1**, 64.3, 0.76
PM: l_1, D_1, M_o	12.7, **17.78**, 1.34T	12.7, **19.05**, 1.34 T
EM: l_2, D_2, D_3,	**25.4, 19.05, 9.525**	**19.05, 20.32, 7.62**
N, I	1040, 4 A	1040, 4 A

Geometrical dimensions are in mm; N = # of turns

Table 3.9 DMP Parameters of PM (n = 6)

Parameters	Design A ($k = 2$)	Design B ($k = 1$)
m_j ($\times 10^{-5}$), A m	4.95/0.1/12.59	−4.72/18.08
$\rho/(D_1/2)$	0.5	0.75
d/l	0.5162	0.3028
% error	3.3	3.6

3.4 Image Method with DMP Models

Table 3.10 Equivalent DMP parameters of ePM (n = 6)

Parameters	Design A (k = 1)	Design B (k = 1)
m_j ($\times 10^{-8}$), A m	1.48/10.83	0.56/5.48
d/l	0.9165	0.9501
% error	5.7	6.73

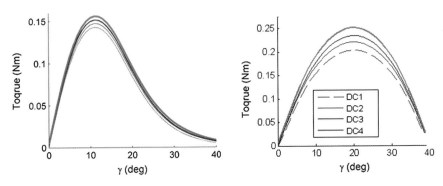

Fig. 3.21 Effect of iron boundaries on torque

Table 3.11 % increase in max. Torque relative to DC1

Designs	DC2 (iron rotor)	DC3 (iron stator)	DC4 (both)
Design A (%)	3.2	6.0	9.2
Design B (%)	8.68	15.15	23.73

$$\% \, \text{error} = 100 \times \int_z |\Phi(z) - \Phi_A(z)| dz / \int_z |\Phi_A(z)| dz$$

In each design, the torque is computed for four design configurations DC1 to DC4 and plotted as a function of the separation angle γ (between the magnetization axes of the PM and EM) in Fig. 3.21. In addition, the percentage increase in the maximum torques of DC2, DC3 and DC4 relative to DC1 are compared in Table 3.11.

Some observations are summarized as follows:

1. In Design A, the combined rotor/stator irons (DC4) contribute to 9.2% increase in the maximum torque; two-thirds are from the iron stator shell while one-third is from the iron rotor. The torque results are consistent with the predicted magnetic field distributions and can be explained with the aid of Fig. 3.18 as follows. Because of the infinitely high permeability of the iron, the magnetic

Fig. 3.22 Magnetic field and torque comparison

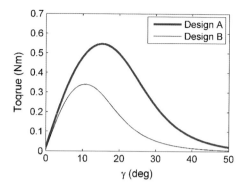

fluxes go through the shortest path into the iron boundary perpendicularly. The shorter flux path implies higher magnetic field intensity and thus has a direct effect on increasing the force acting on the stator EM. In DC1, the path lines between the two PMs in the r_s region are much longer than those in the r_r region. As a result, the iron stator plays a more significant role in shortening the path lines than the iron rotor.

2. The image method with DMP model can be used to investigate the effect of pole designs *with iron boundaries* on the electromagnetic torque of a spherical wheel motor (SWM) [1]. Figure 3.22 compares the generated torque using the two different PM/EM pole designs (Tables 3.8 and 3.9) for the structure shown in Fig. 3.20, where iron conductors were used in both the rotor and stator. The results show that Design B generates 58% larger maximum torque than Design A; both magnetic and mechanical structures contribute to this increase. Mechanically, Design B has a larger moment arm r_r while magnetically Design B has a smaller (r_s-r_r) air region between the rotor and stator reducing magnetic flux leakages in the (r_s-r_r) region. In addition, the shortened EM which has an increased width for the same number of turns results in more current-conducting windings in the stronger magnetic fields.

3.5 Illustrative Numerical Simulations for PMSM Design

With EMs and PMs modeled as DMP, the dipole force model is an efficient way to compute the magnetic force in 3D space for the design of an electromagnetic system, especially for PMSMs involving a large number of EMs and PMs. As an illustrative example, Fig. 3.23 shows the CAD model of a ball-joint-like orientation stage operated on the principle of a PMSM where the PMs are housed in the socket-like rotor assembly. In the figure, the stator electromagnets are air-cored; and the structure (except PMs) is non-magnetic. Supported on a bearing, the rotor is concentric with the stator; thus, the system has three DOF.

3.5 Illustrative Numerical Simulations for PMSM Design

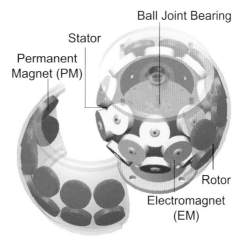

Fig. 3.23 CAD model of a PMSM

The coordinate systems are defined in (a), where XYZ is the stator frame (stationary); xyz is the rotor frame; the Euler angles (roll-pitch-yaw) (α, β, γ) describes the rotor orientation in as:

$$\mathbf{q} = [\alpha, \beta, \gamma]^T \qquad (3.49)$$

The locations of the PMs and EMs as well as the magnetic sensors for measuring the MFD are defined with spherical coordinates. As shown in Fig. 3.24b, $\theta, \phi, r\, (\Theta, \Phi, R)$ represent the spherical coordinates in xyz (XYZ) frame. The magnetization axes of the PMs or EMs can be characterized by a vector pointing from the origin to the center of each PM and EM. The centroids are defined in terms of spherical coordinates (as shown in Fig. 3.24b) in rotor frame (for PMs) and stator frame (for EMs) respectively, which have the following forms:

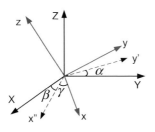

(a) Stator and rotor coordinates and orientation (y' and x'' are intermediate axes)

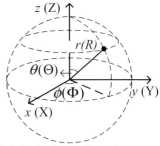

(b) Spherical coordinate in rotor (stator) frame

Fig. 3.24 Coordinate systems of PMSM

Table 3.12 Locations of PMs and EMs

	PM (in xyz)		EM (in XYZ)		
Index	1–12	13–24	1–8	9–16	17–24
$\theta(\Theta)$ (deg)	105	75	116	64	0
$\phi(\Phi)$ (deg)	$30(j-1)$	$30(j-13)$	$45(j-1)$	$45(j-9)$	$45(j-17)+22.5$

$R_{PM} = 67.9$ mm, $R_{EM} = 56.8$ mm, $R_S = 56.4$ mm

Table 3.13 Current input configuration of the EMs

$i_1 = i_{13} = u_1$	$i_5 = i_9 = u_5$	$i_{17} = i_{21} = u_9$
$i_2 = i_{14} = u_2$	$i_6 = i_{10} = u_6$	$i_{18} = i_{22} = u_{10}$
$i_3 = i_{15} = u_3$	$i_7 = i_{11} = u_7$	$i_{19} = i_{23} = u_{11}$
$i_4 = i_{16} = u_4$	$i_8 = i_{12} = u_8$	$i_{20} = i_{24} = u_{12}$

$$\mathbf{C}_{PMi} = R_{PM}[\cos\phi_i \sin\theta_i \quad \sin\phi_i \sin\theta_i \quad \cos\theta_i]^T \quad (3.50a)$$

$$\mathbf{C}_{EMj} = R_{EM}[\cos\Phi_j \sin\Theta_j \quad \sin\Phi_j \sin\Theta_j \quad \cos\Theta_j]^T \quad (3.50b)$$

where i and j are the indices of the PMs and EMs. The parameters are given in Table 3.12. It is worth noting that the adjacent PMs have alternating magnetizations. Due to the symmetric configuration of the rotor PMs, the 24 EMs are grouped in series into pairs leading to a total of 12 electrical inputs (as listed in Table 3.13), which are placed symmetrically about the motor center. The operating range of this design is (Tables 3.12 and 3.13):

$$-22.5° \leq \alpha, \beta \leq 22.5° \quad \text{and} \quad -\infty \leq \gamma < +\infty \quad (3.51)$$

The torque model of the PMSM with linear magnetic properties has the form:

$$\mathbf{T} = [T_X \quad T_Y \quad T_Z]^T = [\mathbf{K}]\mathbf{I} \quad (3.52)$$

where

$$\mathbf{K}(\in \mathbb{R}^{3 \times m_s}) = \begin{bmatrix} \vec{K}_1 & \cdots & \vec{K}_p & \cdots & \vec{K}_{m_s} \end{bmatrix} \quad (3.52a)$$

$$\mathbf{I} = [I_1 \quad \cdots \quad I_p \quad \cdots \quad I_{m_s}]^T \quad (3.52b)$$

I_p is the current input to the pth EM; and m_s is the total number of EMs. In (3.52a), the torque characteristic vector ($\vec{K}_p \in \mathbb{R}^{3 \times 1}$, contributed by I_p to the whole rotor) at each orientation can be derived using the dipole force equation as

3.5 Illustrative Numerical Simulations for PMSM Design

$$\vec{K}_p = \frac{\mu_0}{4\pi} \sum_{i=1}^{n_r} m_{r_i} \sum_{j=1}^{n_p} m_{s_j} \left[\left(\mathbf{R}_{s_{j+}r_{i+}} - \mathbf{R}_{s_{j-}r_{i+}} \right) \times \mathbf{R}_{r_{i+}} - \left(\mathbf{R}_{s_{j+}r_{i-}} - \mathbf{R}_{s_{j-}r_{i-}} \right) \times \mathbf{R}_{r_{i-}} \right]$$
(3.53)

where n_p (or n_r) is the number of dipoles for each EM (or PM).

Inverse Torque Model
Since the three-DOF orientation stage has more current inputs than its mechanical DOF, the optimal current input vector for a given torque is found by minimizing the input energy consumption subject to the required torque constraint. Provided that the input currents are kept within limits, the optimal current input vector can be solved using Lagrange multipliers. For a system where m_s electromagnets are grouped into m independent inputs, the optimal solution for $\mathbf{u} = \begin{bmatrix} u_1 & \cdots & u_p & \cdots & u_m \end{bmatrix}^T$ can be written in closed-form:

$$\mathbf{u} = [\mathbf{K}]^T \left([\mathbf{K}][\mathbf{K}]^T \right)^{-1} \mathbf{T}$$
(3.54)

Simulation results are given here to illustrate the effects of pole sizes on the magnetic torque and the inverse torque model of the orientation stage. In addition, the design of a weight compensating regulator is also illustrated based on the dipole force model and image method.

3.5.1 Pole Pair Design

Effect of pole-shape and design configuration
The geometry and layout of the PM's have a significant influence on the magnetic torque of a PM-based actuator. This example illustrates the use of DMP models to investigate the effect of pole-shapes on the magnetic torque of a spherical motor. The PM pole-shapes used in the following designs are considered:

Design A consists of 2 rows of 8 cylindrical PM's ($\gamma = 1$).
Design B uses a row of 8 assemblies of 5 cylindrical PM's with $\gamma > 1$.
Design C is similar to *Design B* but the customized PM's are used as rotor poles.

The net magnetic torques per unit magnet-volume are compared for a given rotor radius under the same influence of the stator electromagnets. Detailed geometries of the three PM pole-shapes are compared in Fig. 3.25, where the **bold arrow** indicates the polarity of the PM. The parameters used in the simulation are summarized in Table 3.14.

The simulated magnetic flux and potential lines are plotted in Fig. 3.26 where the potential and flux lines are orthogonal as expected. In Fig. 3.26a, the magnetic fields of the three different PM designs are compared in the left column. Unlike

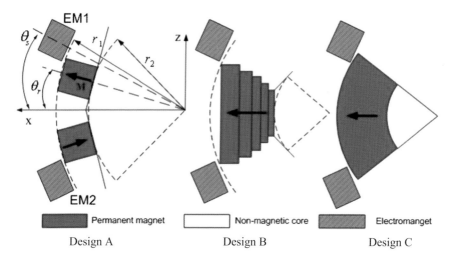

Fig. 3.25 PM pole-shape designs

Table 3.14 Simulation parameters

Common parameters				
Rotor radius (mm)	Stator EM ($\theta_s = 26°$)			Air gap (mm)
	$(2a_o) \times (2a_i) \times \ell$ (mm)[a]	# of turns	Current (A)	
$r_1 = 37.5$	$19.05 \times 9.53 \times 25.4$	1050	±1	0.5
PM pole designs				
	$\mu_0 M_0$ (T)	PM pole shape (mm)		Volume (cm³)
Design A	1.27	$2a = \ell = 12.7$, $\theta_r = 20°$		3.22
Design B	1.27	$2a \times \ell$: 25×10, 20×5, 16×6, 12×3, 8×3		8.2
Design C	0.62	Figure 3.6 ($r_o = 46.5$, $r_i = 23$, $\phi_o = 70°$, $\theta_o = 40°$)		23.6

a_o and a_i represent the outer and inner radii of the EM
a is the outer radius of the PM
ℓ is the length of the EM/PM

Designs B and *C* where only one row of PM's is used, a significant portion of the flux lines in *Design A* forms a closed path between two PM's. Once the magnetic field of the PM's is found the force acting on the current-carrying conductors can be calculated using the Lorentz force equation. Figure 3.27 compares the torque per unit volume of the three designs. The calculation in Fig. 3.27 uses the ESL model with the magnetic field given in Fig. 3.26a. In calculating the torques, ±1A current profiles in Fig. 3.27a, b are given to the EM's such that a positive torque in +y-direction is generated.

3.5 Illustrative Numerical Simulations for PMSM Design

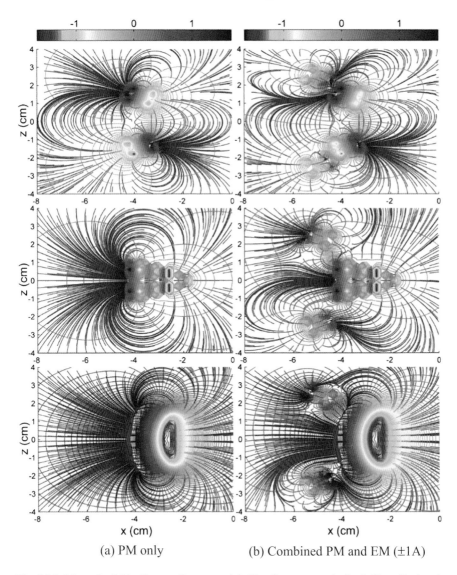

(a) PM only (b) Combined PM and EM (±1A)

Fig. 3.26 Magnetic fields (Orange line: potential; blue lines: magnetic flux) (Top: Design A; middle: Design B; bottom: Design C)

Figure 3.27c shows that *Design A* offers the largest electromagnetic energy to mechanical energy (area under the torque-displacement curve) conversion and that *Design C* has the smallest torque-to-volume ratio. These observations can be explained by comparing the magnetic fields of the designs. As an illustration, snap-shots of the combined (PM and EM) fields for the three designs are compared in Fig. 3.26b, where ±1 A of current is given to the pair of EM's such that a magnetic torque is generated in +y-direction. In other words, the upper EM is

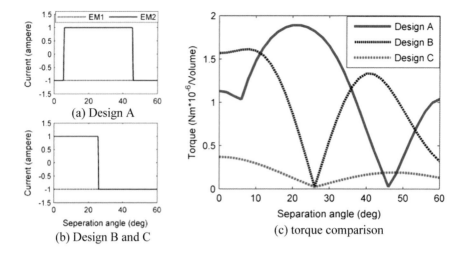

Fig. 3.27 Comparison of torque/volume

repulsive while the lower EM is attractive. The comparison shows that *Design A* has significantly less leakage fluxes in the attractive PM or EM, and less attractive fluxes in the repulsive EM than *Designs B* and *C*. The leakage fluxes in the attractive PM or EM are considered losses as they do not contributes to mechanical torques. Due to the large exposed surfaces in *Designs B* and *C*, a relatively strong closed path of magnetic flux is formed between the repulsive EM and the single PM, which would produce an opposing torque, and thus reduce the net torque. Also, significantly large leakage fluxes from the customized magnet (region between the two EM's) can be seen in *Design C* indicating PM is oversized for the specified EM's. As illustrated in Fig. 3.26b, the closed-form solution of the DMP models can offer an inexpensive means to visualize and analyze the effect of the EM fields on the leakage and unexpected flux paths that have significant influences on the magnetic torque.

Effects of pole size
With EMs and PMs modeled as DMP, the dipole force model is an efficient way to compute the magnetic for the design of electromagnetic systems that involves a large number of EMs and PMs.

Observations in Fig. 3.9 suggest that both small a_r and L (for a given a_o) have a significant effect on the increase in the z-component magnetic fluxes, and hence on the compact design of a spherical motor. The effect can be illustrated with the example in Fig. 3.28 and Table 3.15, where two pole sizes of a PMSM are compared.

Similar to the force model (3.37), the magnetic torque between EMs and PMs can be computed using dipole force model with the DMP models of EMs and PMs, which has the form:

3.5 Illustrative Numerical Simulations for PMSM Design

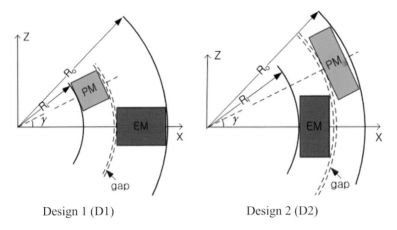

Design 1 (D1) Design 2 (D2)

Fig. 3.28 Comparison of design parameters (R_o = 76.2 mm)

Table 3.15 Parameters defining stator and rotor poles

	PM: ⟵ $2a_o$ ⟶ ↑M ↕l	EM: ⟵$2a_o$⟶⟵$2a_i$⟶ Core ↕l
	Design 1 (D1)	Design 2 (D2)
	R_i = 37.5 mm	R_i = 52.75 mm
PM	a_o = 6.35 mm, L = 1, $\mu_o M_o$ = 1.27 T	a_o = 15.875 mm, L = 0.2, $\mu_o M_o$ = 1.27 T
DMP$_{PM}$	$n = 2, k = 6, \bar{l}/l = 0.7519$ m_i (μA/m): 10.64, 1.68, 37.7	$n = 10, k = 4, \bar{l}/l = 0.3$ m_i(μA/m): 33.5, 24.5, 57.6, 52.0, 276.1
EM	a_o = 9.53 mm, a_r = 0.5, L = 1.33, # of turns = 1050	a_o = 15.88 mm, a_r = 0.3, L = 0.3, # of turns = 1050
DMP$_{EM}$	$n = 12, k = 4, \bar{l}/l = 0.807$ m_i (μA/m): −0.152, 0.448, 0.395, 0.515, 0.0563	$n = 16, k = 6, \bar{l}/l = 0.442$ m_i(μA/m): 1.476, 0.547, 1.618, 1.644, 1.654, 1.325, 0.592
Common parameters: 29AWG, I = 1 A, gap = 0.5 mm, R_o = 76.2 mm		

$$\mathbf{T} = \frac{J\mu_0}{4\pi} \sum_{i=1}^{n_r} m_{r_i} \sum_{j=1}^{n_s} m_{s_j} \left[\left(\mathbf{R}_{s_j+r_{i+}} - \mathbf{R}_{s_j-r_{i+}} \right) \times \mathbf{R}_{r_{i+}} - \left(\mathbf{R}_{s_j+r_{i-}} - \mathbf{R}_{s_j-r_{i-}} \right) \times \mathbf{R}_{r_{i-}} \right]$$

(3.55)

Design 1 (D1) simulates the torque between the rotor PM and stator EM of the SWM where $L \geq 1$ while Design 2 (D2) models that with the same outer radius R_o = 76.2 mm. In D2, both the PM and EM have a much smaller L of 0.2 and 0.3

Fig. 3.29 Effect of pole geometries on actuator torque

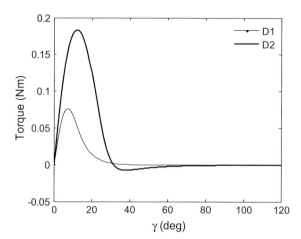

respectively and as a result, the rotor PM (embedded in the "socket") has a 1.4 time larger rotational radius than that of D1. The EM in Table 3.4 is used for D2 and repeated here for ease of comparison.

The effects of the pole size on the magnetic torque are compared in Fig. 3.29 that plots the torque as a function of γ (the separation angle between the magnetization axes of PM and EM). As compared to D1 in Fig. 3.29, D2 offers 2.4 times higher maximum torque and converts 3.6 times more mechanical energy (represented by the area under the torque–displacement curve).

3.5.2 Static Loading Investigation

When the table is loaded (such as a workpiece), the rotor of the three-DOF orientation stage is subjected to an external torque T_{ext} (Fig. 3.30), where the center of gravity coincides with the rotation center.

$$T_{ext} = r \times m_{load} g \quad (3.56)$$

Fig. 3.30 Schematic of the external loading

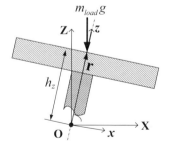

3.5 Illustrative Numerical Simulations for PMSM Design

Table 3.16 Simulation parameters

m_{load} (kg)	h_z (mm)	Rotor mass (kg)	Moment of inertia (kg-m^2)
8	64.8	2.03	$I_{zz} = 7.97 \times 10^{-3}$, $I_{xx} = I_{yy} = 5.89 \times 10^{-3}$

Statically, the torque acting on the rotor is equal to the external torque. The interest here is to simulate the maximum current inputs required meeting a specified torque over the entire operating range given in (3.56).

The required current inputs for any desired torque can be computed using inverse torque model (3.54). The parameters used in simulating the inverse torque model with the component $\mathbf{K}_j$ given by (3.53) are based on Fig. 3.23 and D2 in Table 3.16. Figure 3.31 shows the current profiles of each of the current inputs required maintaining the external torque. Each point represents the maximum current magnitude at the orientation (α, β, $0 \leq \gamma \leq 360°$). Except near the boundary, most of the required currents are within 3A. The statistics of the EM required inputs are summarized in Table 3.17 suggesting that the maximum current required is less 3.4A for the specified load (and rotor weight) of 10 kg.

3.5.3 Weight-Compensating Regulator

The PMSM behaves like an inverted pendulum particularly when the rotor carries a weight. It is desired to develop a method to passively compensate for the static weight (of the rotor and load) such that the rotor is angularly supported against gravity, and maintained at its equilibrium operating point. A common method is to use a counterweight which increases the moment of inertia. An effective alternative is to use repulsive magnetic fields for passive compensation, which has the advantage of quick response. In addition, magnetic fields can be inexpensively measured (using sensors such as Hall-effect) in real-time for position and torque information. However, the advantages of PM-based passive compensation are underexploited as the restoring torque between PMs is generally a nonlinear function of both position and magnetic fields. To develop a method for designing a PM-based passive compensation, a PM-based weight-compensating regulator (WCR) has been developed for the prototype three-DOF PMSM to angularly support its rotor against gravity by means of magnetic repulsion using two circular rings of PMs indicated as RI and RII in Fig. 3.32. RI (fixed with rotor) and RII (fixed with stator), each of which consists of an isolating iron plate and 24 cylindrical PMs equally spaced on the rotor and stator circumferentially respectively, are designed to have like polarities facing each other. Figure 3.33a schematically illustrates the WCR. The centroids of the PMs are defined by (3.57) with cylindrical coordinates in their respective rotor and stator frames where the parameters are given in Table 3.18:

$$\vec{C} = [r\cos\theta_p \quad r\sin\theta_p \quad -h]^{\mathrm{T}} \quad \text{where} \quad \theta_p = \pi p/12, \quad p = 0, 1, \ldots, 23 \quad (3.57)$$

The restoring torque due to the repulsive magnetic force of the WCR can be computed using dipole force model along with image method using the following two steps with the schematic illustrated in Fig. 3.33b:

SI: The effects of the iron plates in the torque computation can be taken into consideration by replacing the iron boundaries with images of the PMs. As shown in Fig. 3.33b, the images of the PMs that are symmetrical about BI and BII are added. The net torque between RI and RII is equivalent to the interactions between the PMs in RI (including their images) and the PMs in RII (including the images).

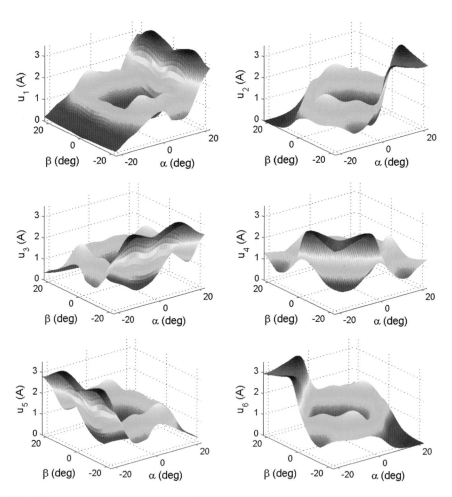

Fig. 3.31 Current inputs in each stator EM

3.5 Illustrative Numerical Simulations for PMSM Design

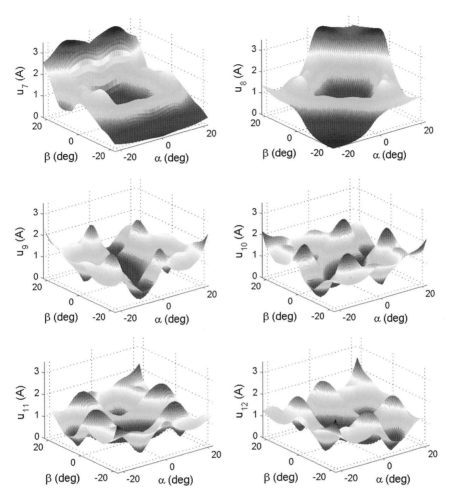

Fig. 3.31 (continued)

Table 3.17 Statistics of current magnitudes (*unit*: A)

u_i	1	2	3	4	5	6	7	8	9	10	11	12
Max	3.34	3.34	3.35	3.35	3.35	3.35	3.35	3.35	2.45	2.18	2.18	2.45
Mean	1.31	1.28	1.30	1.28	1.31	1.28	1.30	1.28	1.06	1.05	1.05	1.06

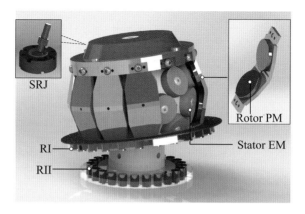

Fig. 3.32 CAD model of PMSM

Table 3.18 WCR parameters

Rings	$h_1 = 33.33$, $r_1 = 73.66$, $h_2 = 12.07$, $r_2 = 49.53$
PMs in RI	$a_o = 9.53$, $l = 9.53$, $\mu_0 M_0 = 1.19$ T
	DMP: $n = 6$, $k = 1$, $\bar{l}/l = 0.417$, $m = [-22.7, 39.2]$ μA/m
PMs in RII	$a_o = 9.53$, $l = 12.70$, $\mu_0 M_0 = 1.19$ T
	DMP: $n = 6$, $k = 1$, $\bar{l}/l = 0.717$, $m = [-1.34, 20.3]$ μA/m

Units in mm otherwise denoted, $\mu_0 = 1.26 \times 10^{-6}$ H/m

SII: The PMs (as well as the images) can be modeled with distributed DMP model and the torque can be then computed using dipole force methods in closed-form with the DMP parameters of the PMs in the WCR given in Table 3.18. The detailed processes for obtaining the DMP parameters of the PMs and calculating magnetic forces with DMP parameters can be found in Sects. 3.1 and 3.3 respectively.

Figure 3.33c shows the computed restoring torque against the rotor inclination angle and it can be seen that the WCR functions as a nonlinear spring tending to return the rotor to its equilibrium (z and Z axes align) when there is no current input.

3.5 Illustrative Numerical Simulations for PMSM Design

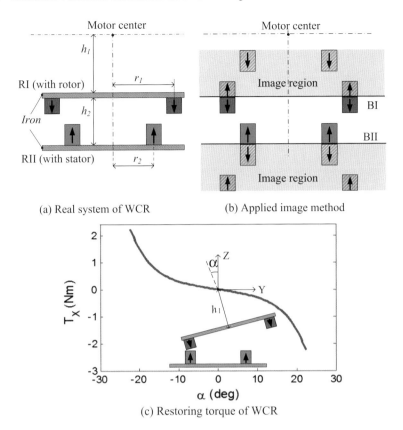

Fig. 3.33 Schematic of WCR and restoring torque (arrows represent the magnetizations; shaded PMs represent "images")

Appendix

The numerical values as shown in Figs. 3.14 and 3.15 are provided in Tables 3.19, 3.20, 3.21, 3.22 and 3.23 for reference.

Table 3.19 Numerical values in Fig. 3.14 (1st row, left column)

$2z/l$	B_Z (10^{-3} T)		
	ML	ESL	DMP_{EM}
1	1.842	2.175	1.863
1.262	1.648	1.951	1.653
1.525	1.447	1.717	1.432
1.787	1.251	1.489	1.221
2.05	1.070	1.277	1.031
2.312	0.910	1.088	0.867
2.575	0.770	0.923	0.728
2.837	0.652	0.782	0.611

Table 3.20 Numerical values in Fig. 3.14 (2nd row, left column)

r/a_0	B_Z/B_{Z0} ($B_{Z0} = 1.3 \times 10^{-3}$ T)		
	ML	ESL	DMP_{EM}
0	1	1	1
0.1918	1.004	1.004	1.003
0.4038	1.009	1.012	1.009
0.5048	1	1.007	1.003
0.6058	0.965	0.980	0.977
0.7067	0.875	0.908	0.897
0.8077	0.665	0.745	0.675
0.9086	0.340	0.331	0.337
1.01	0.063	0.034	0.086
1.1	−0.043	−0.050	−0.028
1.201	−0.083	−0.086	−0.078
1.302	−0.096	−0.099	−0.093
1.403	−0.097	−0.099	−0.093
1.504	−0.092	−0.094	−0.087
1.605	−0.084	−0.086	−0.079
1.706	−0.076	−0.077	−0.070
1.807	−0.067	−0.069	−0.062
1.908	−0.060	−0.061	−0.055
1.999	−0.054	−0.055	−0.049

Appendix

Table 3.21 Numerical values in Fig. 3.14 (3rd row, left column)

r/a_0	B_r (10^{-3} T)		
	ML	ESL	DMPEM
0	0	0	0
0.101	0.098	0.113	0.112
0.2019	0.203	0.234	0.221
0.3029	0.322	0.371	0.349
0.4038	0.463	0.532	0.501
0.5048	0.635	0.729	0.677
0.6058	0.855	0.976	0.895
0.7067	1.184	1.312	1.207
0.8077	1.802	1.979	1.674
0.838	1.920	2.389	1.738
0.8582	1.966	2.763	1.736
0.8884	1.982	3.231	1.687
0.9187	1.955	2.872	1.601
0.949	1.857	2.265	1.487
1.01	1.427	1.533	1.220
1.1	0.886	1.020	0.871
1.201	0.606	0.716	0.608
1.302	0.440	0.532	0.436
1.403	0.328	0.391	0.320
1.504	0.249	0.297	0.240
1.605	0.192	0.229	0.183
1.706	0.150	0.179	0.141
1.807	0.118	0.141	0.111
1.908	0.095	0.113	0.088
1.999	0.078	0.094	0.073

Table 3.22 Numerical values in Fig. 3.15 (1st row, left column)

Tangential displacement (mm)	Force (9.81×10^{-3} N)			
	Exp [11]	I (Maxwell stress tensor)	II (Lorentz force)	III (Dipole force)
0	0	0	0	0
0.020	0.014	0.010	0.013	0.013
0.040	0.031	0.024	0.026	0.025
0.060	0.044	0.038	0.039	0.038
0.080	0.059	0.051	0.052	0.050
0.100	0.074	0.065	0.065	0.061
0.120	0.087	0.079	0.078	0.075
0.140	0.103	0.092	0.091	0.088

(continued)

Table 3.22 (continued)

Tangential displacement (mm)	Force (9.81 × 10⁻³ N)			
	Exp [11]	I (Maxwell stress tensor)	II (Lorentz force)	III (Dipole force)
0.160	0.117	0.105	0.104	0.101
0.180	0.130	0.118	0.117	0.114
0.200	0.144	0.131	0.130	0.127
0.220	0.158	0.144	0.143	0.140
0.240	0.171	0.156	0.155	0.153
0.260	0.185	0.168	0.168	0.166
0.280	0.197	0.180	0.181	0.179
0.300	0.210	0.192	0.193	0.193
0.320	0.223	0.203	0.206	0.207
0.340	0.233	0.214	0.218	0.221
0.360	0.247	0.225	0.231	0.235
0.380	0.259	0.235	0.243	0.249
0.400	0.270	0.245	0.255	0.263
0.420	0.281	0.255	0.267	0.278
0.440	0.293	0.265	0.279	0.292
0.460	0.304	0.274	0.291	0.307
0.480	0.315	0.283	0.302	0.323
0.500	0.326	0.291	0.314	0.338
0.508	0.330	0.297	0.318	0.344
0.520	0.336	0.299	0.325	0.353

Table 3.23 Numerical values in Fig. 3.15 (2nd row, left column)

Axial displacement (mm)	Force (9.81 × 10⁻³ N)			
	Exp [11]	I (Maxwell stress tensor)	II (Lorentz force)	III (Dipole force)
0	1.462	1.570	1.581	1.394
0.014	1.440	1.536	1.558	1.379
0.034	1.409	1.489	1.526	1.357
0.054	1.382	1.444	1.494	1.336
0.074	1.354	1.401	1.463	1.314
0.094	1.326	1.360	1.432	1.293
0.114	1.301	1.321	1.402	1.271
0.134	1.273	1.284	1.373	1.250
0.154	1.248	1.249	1.344	1.229
0.174	1.223	1.215	1.316	1.208

(continued)

Table 3.23 (continued)

Axial displacement (mm)	Force (9.81 × 10^{-3} N)			
	Exp [11]	I (Maxwell stress tensor)	II (Lorentz force)	III (Dipole force)
0.194	1.198	1.182	1.289	1.187
0.214	1.174	1.150	1.262	1.167
0.234	1.150	1.120	1.236	1.146
0.254	1.129	1.092	1.211	1.126
0.274	1.106	1.065	1.186	1.106
0.294	1.084	1.039	1.161	1.087
0.314	1.063	1.014	1.137	1.067
0.334	1.044	0.990	1.114	1.048
0.354	1.025	0.966	1.091	1.029
0.374	1.004	0.943	1.069	1.010
0.394	0.986	0.921	1.047	0.992
0.414	0.967	0.899	1.026	0.974
0.434	0.951	0.879	1.005	0.956
0.454	0.934	0.859	0.985	0.938
0.474	0.917	0.840	0.965	0.921
0.494	0.901	0.821	0.946	0.903
0.508	0.8898	0.810	0.933	0.892
0.514	0.885	0.804	0.927	0.887

References

1. K.-M. Lee, H. Son, Distributed multipole model for design of permanent-magnet-based actuators. IEEE Trans. Magn. **43**, 3904–3913 (2007)
2. H. Son, K.-M. Lee, Distributed multipole models for design and control of PM actuators and sensors. IEEE-ASME Trans. Mechatron. **13**, 228–238 (2008)
3. W.S. Bennett, Basic sources of electric and magnetic fields newly examined. IEEE Antennas Propag. Mag., 31–35 (2001)
4. D.J. Craik, Magnetostatics of axially symmetric structure. J. Phys. **7**(11), 1566 (1974)
5. J.D. Jackson, *Classical electrodynamics* (Wiley, New York, 1999)
6. Y. Liang, I.M. Chen, G. Yang, L. Wei, K.-M. Lee, Analytical and experimental investigation on the magnetic field and torque of a permanent magnet spherical actuator. IEEE/ASME Trans. Mechatron., 409–419 (2006)
7. Y. Liang, Modeling and design of a three degree-of-freedom permanent magnet spherical actuator, Ph.D. thesis, Nanyang Technological University (Singapore), 2005
8. K.M. Lee, K. Bai, J. Lim, Dipole models for forward/inverse torque computation of a spherical motor. IEEE/ASME Trans. Mechatron. **14**, 46–54 (2009)
9. K.-M. Lee, H. Son, Equivalent voice-coil models for real-time computation in electromagnetic actuation and sensor applications, in *Proceedings of AIM2007*, September 4–7, 2007, ETH Zürich, Switzerland
10. H.A. Haus, J.R. Melcher, Electromagnetic fields and energy [Online]. Available: https://ocw.mit.edu/resources/res-6-001-electromagnetic-fields-and-energy-spring-2008/

11. N.I.J.P.A. Bastos, Forces in permanent magnets. Team workshop problems 23 [Online]. Available: http://www.compumag.org/jsite/team.html
12. K.-M. Lee, H. Son, K. Bai, Image method with distributed multipole models for analyzing permanent-magnet-based electromagnetic actuators, in *Proceedings of the ASME Dynamic Systems and Control Conference* (2009), pp. 791–797
13. H. Son, K. Bai, J. Lim, K.M. Lee, Design of multi-DOF electromagnetic actuators using distributed multipole models and image method. Int. J. Appl. Electromagnet. Mech. **34**(3), 195–210 (2010)
14. K.-M. Lee, R.A. Sosseh, Z. Wei, Effects of the torque model on the control of a VR spherical motor. IFAC J. Control Eng. Pract. **12**(11), 1437–1449 (2004)
15. K.-M. Lee, J. Joni, H. Son, Design method for prototyping a cost-effective variable-reluctance spherical motor (VRSM), in *Proceedings of IEEE Conference on Robotics, Automation and Mechatronics*, Singapore, December 2004

Chapter 4
PMSM Force/Torque Model for Real-Time Control

In Chap. 3, the DMP-based force/torque models was presented as a forward model that solves for the magnetic forces/torques of a PMSM in terms of given design parameters and a set of specified input currents. Once the design parameters are determined, the control of a PMSM requires an inverse magnetic force/torque model to determine the current inputs to generate the desired torque. Different from the DMP-based models in Chap. 3 targeted for reducing computational time for design, analysis and parametric optimization, the solutions to the inverse model of a PMSM must be computed in real-time to allow for instantaneous update of the required magnetic force/torque at any orientation.

In this chapter, a computationally-efficient force/torque model for a PMSM is presented, which describes the torque-to-current relationship in terms of the relative positions among the rotor-PMs and stator-EMs by means of a characteristic kernel function. The kernel function represents the force/torque in an EM-PM pair of a PMSM, through which the torque-to-current relationship can be expressed in closed-form allowing for real-time implementations in a PMSM control system. This method not only can be applied on PMSMs featured with rotational motions, but also can be used as a real-time force/torque computation methods for actuator systems of various forms, such as linear, planar, and z-θ motors.

4.1 Force/Torque Formulation

Consider an electromagnetic motion system consisting of stationary EMs and moving PMs with their centroids locate in two surface domains Γ_S and Γ_R. To allow for a relative motion between the two surfaces, Γ_S and Γ_R are parallel but separated with a geometrical gap g. Figure 4.1 illustrates an EM-PM pair in the motion system with the EM and PM centroids (C_P and C_E) on Γ_S and Γ_R respectively, where (η_E, ζ_E, n_E) and (η_P, ζ_P, n_P) are the local frames of the EM and the PM with n normal to the tangential plane formed by (η, ζ) at their respective

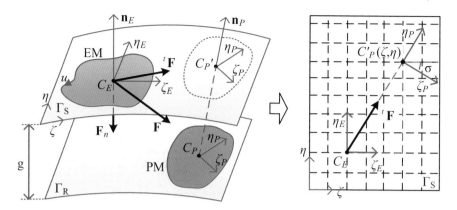

Fig. 4.1 Illustration of an EM-PM pair in motion systems

surface domains. The Lorentz force **F** between the current-carrying EM and the PM can be decomposed into (tangential) $^t\mathbf{F}$ and (normal) $^n\mathbf{F}$ components, where $^n\mathbf{F}$ contributes no motion as the PM moves in the 2-D domain Γ_R. The tangential force $^t\mathbf{F}$ is expressed in terms of the unit vectors ($\vec{e}_{\zeta_E}$ and $\vec{e}_{\eta_E}$) along the local axes (η_E and ζ_E) respectively as follows:

$$^t\mathbf{F} = \left(\mathbf{F} \cdot \vec{e}_{\zeta_E}\right)\vec{e}_{\zeta_E} + \left(\mathbf{F} \cdot \vec{e}_{\eta_E}\right)\vec{e}_{\eta_E} \qquad (4.1)$$

As illustrated in Fig. 4.2, different motion types (depending on the design configurations of the EM-PM pairs) can be expressed in terms of the relative motion between Γ_S and Γ_R.

4.1.1 Magnetic Force/Torque Based on the Kernel Functions

By projecting the PM centroid as well as its axes η_P, ζ_P on Γ_S along the normal direction as shown in Fig. 4.1, the relative position between an EM and a PM in the 3-D space can be characterized by the relative position of the EM centroid C_E and the PM centroid projection C_P' on Γ_S (mapped in a plane as shown in Fig. 4.3), where the actuation force $^t\mathbf{F}$ can be computed using (4.1). As the Lorentz force between the EM (current u) and PM (magnetization m) only depends on their relative position, $^t\mathbf{F}$ can be expressed in terms of the kernel functions $\left(\ell_\eta, \ell_\zeta\right)$ as

$$^t\mathbf{F} = mu\left[\ell_\zeta(\zeta,\eta,\sigma)\vec{e}_\zeta + \ell_\eta(\zeta,\eta,\sigma)\vec{e}_\eta\right] \qquad (4.2)$$

4.1 Force/Torque Formulation

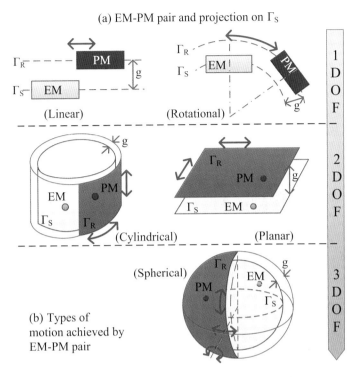

Fig. 4.2 Illustration of EM-PM configurations for different types of motion

Fig. 4.3 Interaction among multiple EMs/PMs

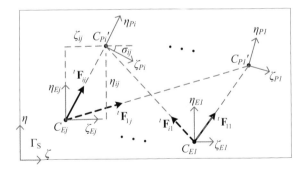

In (4.2), (ζ, η) represent the PM projected-centroid C_P' in the EM frame; and σ represents the rotation angle between the two frames as shown in Fig. 4.1. The kernel functions are found by comparing (4.1) and (4.2):

$$\ell_\zeta(\zeta, \eta, \sigma) = \frac{1}{mu}\left({}^t\mathbf{F} \cdot \vec{e}_{\zeta_E}\right) \tag{4.3a}$$

$$\ell_\eta(\zeta, \eta, \sigma) = \frac{1}{mu}\left({}^t\mathbf{F} \cdot \vec{e}_{\eta_E}\right) \quad (4.3b)$$

For an electromagnetically linear system consisting of N_P identical PMs and N_E identical EMs, the actuation force can be characterized by the interactions among the EMs and PM projections on Γ_S as illustrated in Fig. 4.3. The actuation force applied on the *j*th EM (EM_j) by all PMs can be derived by superimposing the individual forces between EM_j and each PM:

$$\mathbf{{}^tF}_j = \sum_{i=1}^{N_P} \mathbf{{}^tF}_{ij} = u_j \mathbf{K}_j \quad (4.4)$$

where

$$\mathbf{K}_j = \sum_{k=1}^{N_P} m_k \begin{bmatrix} \ell_\zeta(\zeta_{jk}, \eta_{jk}, \sigma_{jk})\vec{e}_{\zeta_{Ej}} \\ + \ell_\eta(\zeta_{jk}, \eta_{jk}, \sigma_{jk})\vec{e}_{\eta_{Ej}} \end{bmatrix} \quad (4.4a)$$

In (4.4a), $(\eta_{jk}, \zeta_{jk}, \sigma_{jk})$ represent the position and orientation of PM_k with respect to the local frame of EM_j; and $\mathbf{K}_j$ is the *characteristic force vector (CFV)* of EM_j. The total actuation force applied on the rotor/mover (reaction force to that applied on the EMs) can be derived as (4.5):

$$\mathbf{F}_A = -\sum_{j=1}^{N_E} \left(\mathbf{{}^tF}_j\right) = [\mathbf{K}]\mathbf{u} \quad (4.5)$$

where

$$[\mathbf{K}] = [\ldots \quad -\mathbf{K}_j \quad \ldots] \quad \text{and} \quad \mathbf{u} = [\ldots \quad u_j \quad \ldots]^T \quad (4.5a,b)$$

For a rotational system, the torque can be computed as (4.6) where s_{Ej} represent the vector from the rotation center to the centroid of EM_j:

$$\mathbf{T}_A = -\sum_{j=1}^{N_E} \left(\mathbf{s}_{Ej} \times \mathbf{{}^tF}_j\right) = [\mathbf{P}]\mathbf{u} \quad (4.6)$$

$$[\mathbf{P}] = [\ldots \quad \mathbf{s}_{Ej} \times \mathbf{K}_j \quad \ldots] \quad (4.6a)$$

With (4.5) and (4.6), the actuation force/torque of a PM-driving system can be represented explicitly in terms of the kernel functions (ℓ_η, ℓ_ζ included in *CFV* $\mathbf{K}_j$) that is identical for all EM-PM pairs in the system.

4.1.2 Simplified Model: Axis-Symmetric EMs/PMs

Axis-symmetric (like cylinder, cone) EMs/PMs are commonly used in electromagnetic systems (especially multi-DOF systems) to generate a symmetric magnetic field distribution. This symmetric property can be used to further simplify the force/torque model (4.5)–(4.6) in terms of the kernel functions (4.3a)–(4.3b). As illustrated in Fig. 4.4, the relative position of the EM-PM pair can be characterized solely by the line connecting their centroids on Γ_S; thus, the actuation force can be expressed as

$$^t\mathbf{F} = mu\ell(\lambda)\vec{e}_C \qquad (4.7)$$

where

$$\lambda = \left| \widehat{C_E C_P}' \right| \qquad (4.7a)$$

and

$$\vec{e}_C = \left(\mathbf{n}_E \times \vec{e}_{C_E C_P'} \right) \times \mathbf{n}_E \qquad (4.7b)$$

In (4.7a) and (4.7b), λ is the length of the line connecting C_E and $C_{P'}$ on Γ_S; $\vec{e}_C$ is the tangent of $\widehat{C_E C_P}'$ at C_E; $\vec{e}_{C_E C_P'}$ is the unit vector from C_E to $C_{P'}$; and $(\mathbf{n}_E, \mathbf{n}_P)$ are the normal vectors of Γ_S at $(C_E, C_{P'})$. As a result, the actuation force can be characterized by a much simpler one-dimensional kernel ℓ that equals to the magnitude of $^t\mathbf{F}$ normalized to mu according to (4.7):

$$\ell(\lambda) = \frac{1}{mu} |^t\mathbf{F}| \qquad (4.8)$$

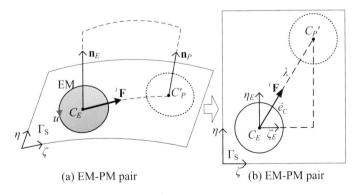

(a) EM-PM pair   (b) EM-PM pair

Fig. 4.4 Interaction between axis-symmetric EM and PM

Therefore, the total force/torque for a driving system consisting of N_P identical PMs and N_E identical EMs can be computed from (4.5) to (4.6) but with a simpler CFV:

$$\mathbf{K}_j = \sum_{k=1}^{N_P} m_k \ell(\lambda_{jk}) \vec{e}_{Cj} \qquad (4.9)$$

In (4.9), λ_{jk} is the length of the line connecting the EM_j centroid and the projection of PM_k centroid on Γ_S; and $\vec{e}_{Cj}$ is the tangent of the connecting line at C_{Ej}.

4.1.3 Inverse Torque Model

With the forward model (4.5)–(4.6) written in standard form $\mathbf{Y} = [\mathbf{X}]\mathbf{u}$, an inverse model that solves for the current inputs for a specified force/torque can be derived with Lagrange multiplier under the condition that the total energy is minimized:

$$\mathbf{u} = [\mathbf{X}]^T \left([\mathbf{X}][\mathbf{X}]^T\right)^{-1} \mathbf{Y} \qquad (4.10)$$

4.2 Numerical Illustrations

The magnetic force/torque model based on the kernel functions are illustrated for EM-PM pair in different geometries and different types of motion. To allow for real-time computation, several fitting methods to characterize the kernel functions in closed form are illustrated.

4.2.1 Axis-Asymmetric EM/PMs

Figure 4.5a illustrates a EM-PM model where the square-PM moves with respect to the stationary square-EM in Cartesian coordinates. Three types of motions, one-DOF, two-DOF and three-DOF, are considered as illustrated in Fig. 4.5b–d respectively. The analytical solutions to the actuation force are numerically derived from the Lorentz force integral Eq. (2.26a), which provide a basis to compute the kernel functions in closed-forms.

As illustrations, different fitting techniques to obtain closed-form approximations of the kernel functions for modeling the characteristic force vectors in the one-DOF, two-DOF and three-DOF motion applications are illustrated in the following subsections. The accuracy and computational-efficiency of the fitting methods are

4.2 Numerical Illustrations

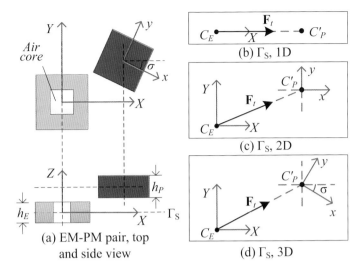

Fig. 4.5 Planar motion with non-axis-symmetric EM/PM

Table 4.1 Simulation parameters

Square EM/PM parameters (mm)			
EM (Air cored)	$a_i = 10$, $a_0 = 30$, thickness = 10, wire dia.: 0.5	PM	$b = 30$, thickness = 6, $\mu_0 M_0 = 1.465$

Fit function properties				
DOF	Fitting parameters	Comp. time (s)		% Error
		Numerical	Fit	
One	$[a_0, \ldots a_5] = [2.25, -2.18, -1.84, 2.19, -0.18, -0.22]$; $[b_0, \ldots b_5] = [0, -2.91, 3.47, 0.33, -0.98, 0.18]$; $\omega = 70.95$	64.6	0.016	0.42
Two	Equation (4.14)	129.2	0.030	3.7
Three	# of hidden layer: 1 # of nodes in the hidden layer: 20	129.2	0.064	7.2

Air gap between EM and PM: $\delta g = 1$ mm

evaluated relative to the analytical solutions in terms of the relative percentage error (RPE) which is computed as the mean absolute error with respect to the peak-to-peak value of the fitted function. The geometrical parameters of the EM/PM and the fitting functions used in the illustrations as well as the computational time and RPEs are summarized in Table 4.1.

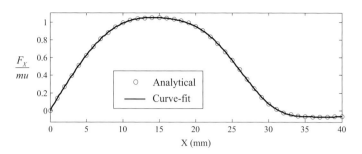

Fig. 4.6 1D force kernel function

1-DOF (Fig. 4.5b)

For a linear motion application, the actuation force model (4.5) reduces to one dimensional characterized by (4.11):

$$F_X = mu\,\ell_X(X) \tag{4.11}$$

Figure 4.6 shows the analytical result of the actuation force along the X-axis normalized to (*mu*). A Fourier function with the form of (4.12) is obtained as a closed-form representation of the kernel function by fitting F_X/mu and the fitted results are compared with the analytical results in Fig. 4.6:

$$\ell_X(X) = \sum_{k=0}^{5} \left[a_j \sin(kcX) + b_j \cos(kcX) \right] \tag{4.12}$$

The fitting parameters, RPE as well as the computational times of the analytical solution and fitting function are listed in Table 4.1. It can be seen that the computational time can be greatly reduced with a fit function with high accuracy.

The CFV of EM_j can be obtained as (4.13) by simplifying (4.4a):

$$K_j(\in \Re^1) = \sum_{k=1}^{N_P} m_k \ell_X(X_{kj}) \vec{e}_{Xj} \tag{4.13}$$

2-DOF (Fig. 4.5c)

When the PM translates in the plane in 2D (without spinning), the 2D kernel functions that characterize the force components acting on the PM take the form:

$$\ell_X(X,Y) = \frac{1}{mu}({}^t\mathbf{F} \cdot \vec{e}_X) \tag{4.13a}$$

$$\ell_Y(X,Y) = \frac{1}{mu}({}^t\mathbf{F} \cdot \vec{e}_Y) \tag{4.13b}$$

4.2 Numerical Illustrations

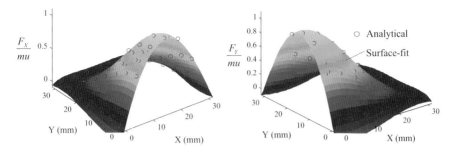

Fig. 4.7 Illustration of fit function for 2D kernel function

Using the analytical solutions computed from (2.26a) shown in Fig. 4.7, the 2D kernel functions are obtained using the 2D fit functions of the form:

$$\ell_{X(Y)}(X,Y) = (1, \eta, \ldots \eta^m) \begin{bmatrix} p_{00}(q_{00}) & \cdots & p_{0n}(q_{0n}) \\ \vdots & \ddots & \vdots \\ p_{n0}(q_{n0}) & \cdots & p_{mn}(q_{mn}) \end{bmatrix} \begin{pmatrix} 1 \\ \vdots \\ \zeta^n \end{pmatrix} \quad (4.14)$$

where

$$[p_{ij}] = [q_{ij}]^{\mathrm{T}} = \begin{bmatrix} -0.0069 & 156.6 & -6676 & 1.26e5 & -4.89e6 & 8.01e7 \\ -2.23 & 959.1 & -7.68e4 & 3.82e6 & -8.17e7 & 0 \\ -90.73 & -5.68e5 & 1.89e7 & -1.84e6 & 0 & 0 \\ 3.16e4 & 1.27e7 & -4.19e8 & 0 & 0 & 0 \\ -1.36e6 & -1.32e6 & 0 & 0 & 0 & 0 \\ 1.59e7 & 0 & 0 & 0 & 0 & 0 \end{bmatrix}$$

The fit function parameters as well as the percentage error of the approximations relative the analytical solutions and the time required for the computation are listed in Table 4.1. With the kernel functions expressed in (4.14), the CFV can be obtained from (4.4a) as:

$$\mathbf{K} = m[\ell_X(X,Y)\vec{e}_X + \ell_Y(X,Y)\vec{e}_Y] \quad (4.15)$$

3-DOF (Fig. 4.5d)

When the PM moves in the plane in a 3D (with spinning) space, the force component functions for the EM-PM characterized by Fig. 4.5d become

$$\ell_X(X,Y,\sigma) = \frac{1}{mu}({}^t\mathbf{F} \cdot \vec{e}_X) \quad (4.16a)$$

$$\ell_Y(X,Y,\sigma) = \frac{1}{mu}({}^t\mathbf{F} \cdot \vec{e}_Y) \quad (4.16b)$$

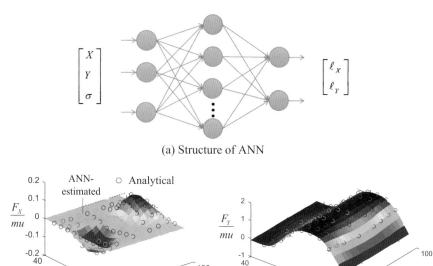

(a) Structure of ANN

(b) Comparison between analytical and ANN-estimated results

Fig. 4.8 Illustration of fit function for 3D kernel function

The force components computed at each (X, Y, σ) are used to obtain a time-efficient fit function. For this, an artificial neural network (ANN) as shown in Fig. 4.8 is used, where the ANN parameters (weights) are trained with the input-output pairs formed by [(X, Y, σ), computed force components]. The ANN parameters, relative percentage error and computational time are listed in Table 4.1.

The CFV can be obtained from (4.4a) as:

$$\mathbf{K} = m[\ell_X(X, Y, \sigma)\vec{e}_X + \ell_Y(X, Y, \sigma)\vec{e}_Y] \qquad (4.17)$$

4.2.2 Axis-Symmetric EM/PM

Figure 4.9 shows two different systems (each consisting of a cylindrical EM/PM pair); and their respective kernel functions are computed for the relative planar (Fig. 4.9a) and spherical (Fig. 4.9b) motions. The relative position of the (EM and PM) centroids is illustrated in Fig. 4.9c.

4.2 Numerical Illustrations

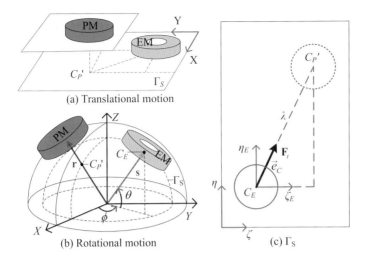

Fig. 4.9 EM-PM interaction with axis-symmetric EM/PM

Translational (Fig. 4.9a)

When the PM moves with 2-DOF on the plane in Fig. 4.9a where the relative positions between EM/PM are illustrated by Fig. 4.9c with ($\zeta = X, \eta = Y$), the kernel functions can be characterized with a 1D function due to symmetry as shown in (4.18):

$$\mathbf{K} = m\ell(\lambda)\vec{e}_C \qquad (4.18)$$

$$\lambda = \left|\mathbf{p}_{C_P'} - \mathbf{p}_{C_E}\right|, \quad \text{and} \quad \vec{e}_C = \left(\mathbf{p}_{C_P'} - \mathbf{p}_{C_E}\right)/\lambda \qquad (4.18\text{a, b})$$

where $\mathbf{p}_{C_P'}$ and $\mathbf{p}_{C_E}$ represent the centroid positions of the EM and PM on Γ_S. Figure 4.10 shows the normalized actuation force as a function of the length of the line segment between C_E and C_P'. A curve-fit function with Fourier form (4.12) was used to approximate the results. The fit-function parameters as well as the relative percentage error and the computational times are listed in Table 4.2.

Rotational (Fig. 4.9b, c)

When the PM rotates in the 3D space as illustrated in Fig. 4.9a, where **r** and **s** represent the position vectors from rotation center to the EM/PM centroids, the relative positions between the EM/PM centroids in the spherical surface Γ_S (defined with θ, ϕ as shown in Fig. 4.9b) can be mapped in a 2D domain as shown in Fig. 4.9c where $\zeta = \theta$ and $\eta = \phi$.

$$\mathbf{K} = m\ell(\lambda)\vec{e}_C \qquad (4.19)$$

$$\lambda = |\mathbf{s}|\rho, \quad \text{where } \rho = \cos^{-1}\left(\frac{\mathbf{r}\cdot\mathbf{s}}{|\mathbf{r}||\mathbf{s}|}\right) \text{ and } \vec{e}_C = \frac{(\mathbf{r}\times\mathbf{s})\times\mathbf{s}}{|(\mathbf{r}\times\mathbf{s})\times\mathbf{s}|} \qquad (4.19\text{a, b, c})$$

Table 4.2 Simulation parameters

Round EM/PM parameters (mm)

EM		PM	
$a_i = 10$, $a_0 = 30$, thickness = 10, wire dia.: 0.5		$b = 30$, thickness = 6, $\mu_0 M_0 = 1.465$	

Fit function properties

Type	Fitting parameters	Comp. time (s)		% Error
		Num.	Fit	
Translation	$[a_0, \ldots a_5] = [0.11, -0.08, -0.14, 0.09, 0.03, -0.01]$; $[b_0, \ldots b_5] = [0, 0.71, 0.31, 0.03, -0.02, -0.02]$; $\omega = 102.3$	71.3	0.016	0.46
Rotation	$[a_0, \ldots a_5] = [-0.0007, 0.02, -0.10, -0.03, 0.07, 0.04]$; $[b_0, \ldots b_5] = [0, 0.74, 0.81, 0.35, 0.05, -0.01]$; $\omega = 74.05$	105.2	0.016	0.45

Air gap between EM and PM: $\delta g = 1$ mm

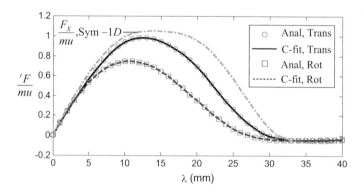

Fig. 4.10 Kernel functions for cylindrical EM/PM

Specifically, the 1D kernel results with cuboid EM/PM as shown in Fig. 4.6 is added in Fig. 4.10 for comparison. It can be seen that for systems consisting of cuboid and cylindrical EMs/PMs that takes up same installation space (as shown in Table 4.1), the maximum FCF with cuboid EM/PM is 5% larger than that with cylindrical EM/PM. Therefore, for a driving system especially multi-DOF motion system, the computational efficiency and thus fast control response can be achieved by using axis-symmetric EMs/PMs and the tradeoff is 5% actuation force. (The areas under the curve are 0.017 and 0.021.)

4.3 Illustrative PMSM Torque Modelling

With the torque model presented in Sect. 4.1, the formulation of the torque-to-current relationship for real-time applications is illustrated for a PMSM (Fig. 4.11a) which is based on the design as shown in Fig. 3.23 with the parameters listed in Table 4.3.

Figure 4.11b shows an EM-PM pair in the PMSM, where both PM and EM are cylindrical and the parameters are given in Table 4.3; XYZ and xyz (sharing a common origin) are the stator reference and rotor local frames respectively; **s** and **r** (with lengths of R_E and R_P) represent the position vectors in XYZ frame from the origin to the geometrical centers of the stator EM and rotor PM; and ρ is the separation angle between **r** and **s**. Figure 4.11b, y' and x'' are two intermediate axes for defining XYZ Euler angles (α, β, γ). Given the PM position vector **p** in xyz coordinates, **r** can be obtained from the rotation matrix [**R**] in (4.20a) and (4.20b) where S and C represent sine and cosine of the Euler angles (subscripts) respectively:

$$\mathbf{r} = [\mathbf{R}]^T \mathbf{p} \qquad (4.20a)$$

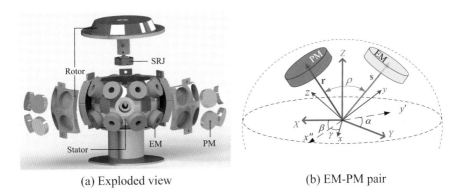

(a) Exploded view (b) EM-PM pair

Fig. 4.11 CAD model of a PMSM

Table 4.3 Parameters of the PMSM

ith PM centroid	Index	(θ, ϕ) in xyz frame		Magnetization, m
	1–12	105°	$(i-1) \times 30°$	$\mu_0 M_0 \, (-1)^i$
	13–24	75°	$(i-13) \times 30°$	$\mu_0 M_0 \, (-1)^{i-1}$
jth EM centroid	Index	(θ, ϕ) location in XYZ frame		
	1–8	−26°	$(j-1) \times 45°$	
	9–16	26°	$(j-9) \times 45°$	

Rotor mass (with counter weight): 1.95 kg, center of mass (h_z): −7.1 mm

where

$$[\mathbf{R}] = \begin{bmatrix} C_\gamma C_\beta & S_\gamma C_\alpha + C_\gamma S_\beta S_\alpha & S_\gamma S_\alpha - C_\gamma S_\beta C_\alpha \\ -S_\gamma C_\beta & C_\gamma C_\alpha - S_\gamma S_\beta S_\alpha & C_\gamma S_\alpha + S_\gamma S_\beta C_\alpha \\ S_\beta & -C_\beta S_\alpha & C_\beta C_\alpha \end{bmatrix} \quad (4.20b)$$

Note that **p** and **s** are constant vectors which are defined in spherical coordinates (with the parameters given in Table 4.2).

For the kernel functions shown in Fig. 4.10 and the fit parameters listed in Table 4.1, the torque characteristic vector (TCV) of the jth EM in stator can be obtained as:

$$\mathbf{K}_j = \sum_{k=1}^{N_P} m_k \ell(\lambda_{jk}) \vec{e}_{Cj} \quad (4.21)$$

where

$$\lambda_{ij} = |\mathbf{s}_j| \rho_{ij}, \quad \text{and } \rho_{ij} = \cos^{-1}\left(\frac{\mathbf{r}_i \cdot \mathbf{s}_j}{|\mathbf{r}_i||\mathbf{s}_j|}\right), \quad \vec{e}_{Cj} = \frac{(\mathbf{r}_i \times \mathbf{s}_j) \times \mathbf{s}_j}{|(\mathbf{r}_i \times \mathbf{s}_j) \times \mathbf{s}_j|} \quad (4.21a, b, c)$$

The total torque applied on the rotor by N_E EMs can be obtained as:

$$\mathbf{T} = [T_X \quad T_Y \quad T_Z]^T = [\mathbf{K}]\mathbf{u} \quad (4.22)$$

where

$$[\mathbf{K}](\in \mathbb{R}^{3 \times m_s}) = [\mathbf{K}_1 \quad \cdots \quad \mathbf{K}_j \quad \cdots \quad \mathbf{K}_{NE}] \quad (4.22a)$$

and

$$\mathbf{u} = [u_1 \quad \cdots \quad u_j \quad \cdots \quad u_{NE}]^T \quad (4.22b)$$

The optimal solution for the current inputs can be written in closed-form:

$$\mathbf{u} = [\mathbf{K}]^T \left([\mathbf{K}][\mathbf{K}]^T\right)^{-1} \mathbf{T} \quad (4.23)$$

(4.22) and (4.23) are referred to here as the forward and inverse torque models respectively.

In order to test the accuracy and the feasibility for real-time applications, the torque model as well as the inverse model has been evaluated on a PMSM test-bed as shown in Fig. 4.12. As shown in the figure, the rotor of the PMSM is pivoted to rotate about a horizontal shaft that is fixed through a ball-bearing mounted on the brackets, and the inclination angle can be acquired with the gyroscope in real-time. A counter-weight is added to the rotor to lower the center of mass of the rotor below the rotation center. In the experiment, a series of desired inclination angles were specified and the current inputs to generate electromagnetic torque to balance the rotor at any inclination angle were computed in real time with the forward and

4.3 Illustrative PMSM Torque Modelling

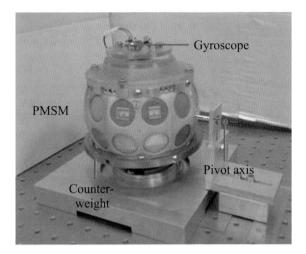

Fig. 4.12 Experimental PMSM setup for torque measurement

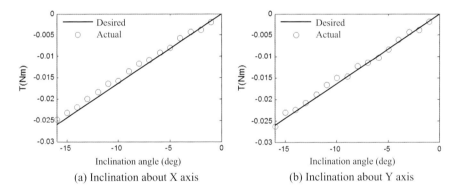

Fig. 4.13 Experimental results

inverse torque models (4.22) and (4.23). At any equilibrium, the actuation torque is assumed to completely support the rotor:

$$T_A = mgh_z \sin \theta \tag{4.24}$$

where θ is the inclination angle, h_z is the distance from the mass center to the rotation center. The values of the parameters are listed in Table 4.3.

The sampling rate of the open-loop control test was 1 ms. The test was performed when the rotor inclined about X and Y axes respectively. The desired torque values at the recorded equilibrium inclination angles are compared with the torque values computed using (4.24) in Fig. 4.13. It can be seen that the actual torques match the desired values very well.

Part II
Sensing Methods

Chapter 5
Field-Based Orientation Sensing

Magnetic sensors are commonly utilized as a media for actuation and sensing. Sensing systems that operate on this principle are able to function in harsh conditions as magnetic fields are invariant to temperature, pressure, radiation and other environmental factors. A multi-sensor approach that capitalizes on the existing magnetic fields in PMSMs to achieve unobtrusive high-accuracy orientation sensing [1, 2] is presented in this chapter.

5.1 Coordinate Systems and Sensor Placement

Figure 5.1 defines the coordinate systems of the stator, rotor and sensors, which are denoted as XYZ (reference), xyz (moving), and $X_pY_pZ_p$ (fixed local) respectively.

The orientation of the XYZ frame is described by a sequence of body-fixed rotations about Z, y and finally z axis by the corresponding angles of ψ, θ and ϕ respectively and this transformation can be represented by

$$[x \quad y \quad z]^T = \Gamma(\mathbf{q})[X \quad Y \quad Z]^T \qquad (5.1)$$

where $\mathbf{q} = [\psi, \theta, \phi]^T$ and Γ is the zyz Euler transformation matrix. O coincides with the center of the spherical bearing.

The stator contains s sensors (blue circles) spaced equally along a circular path C of radius R_s in a plane parallel to the XY plane. This sensor plane is displaced by $-H_s$ along the Z-axis. The pth sensor position in the XYZ plane is

$$\mathbf{S}_p = R_s[\cos((p-1)\psi_s) \quad \sin((p-1)\psi_s) \quad -H_s]^T \qquad (5.2)$$

where $p = 1,2,\ldots,s$; and $\psi_s = 2\pi/s$ is the angular spacing between adjacent sensors. The sensors are orientated such that the sensing X_p and Y_p axis of each sensor are

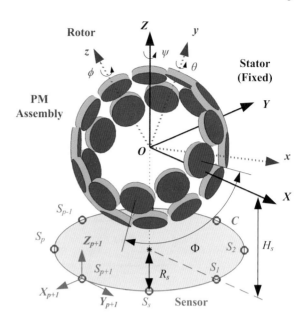

Fig. 5.1 Coordinate systems for rotor, stator and sensors

normal and tangential to **C**. Hence the coordinate transformation from the stator reference frame to the *pth* sensor frame can be described by

$$\begin{bmatrix} X_p & Y_p & Z_p \end{bmatrix}^T = \Gamma_S((p-1)\psi_s)[X - R_s \quad Y \quad Z + H_s]^T \quad (5.3)$$

where Γ_S is the transformation matrix denoting rotation about Z-axis.

5.2 Field Mapping and Segmentation

Due to the alternating PM magnetization, spatial periodicity exists about the z-axis (a 2 × 2 PM configuration constitutes a single spatial period). If the axis of inclination is fixed or known (if $\psi = 0$, the inclination axis will coincide with the stator Y-axis), a 2-D surface field map containing the magnetic flux density measurements of each axis (X_p, Y_p, Z_p) by each sensor at arbitrary inclination (θ) and spin (ϕ) can be constructed. As the range of rotor inclination is limited to $-\theta_{max} \leq \theta \leq \theta_{max}$ due to physical constraints, the domain of this surface map is

$$-\theta_{max} \leq \theta \leq \theta_{max}, -\pi \leq \phi \leq \pi$$

If no symmetry is assumed or present, a typical surface map for the *pth* sensor is shown in Fig. 5.2a. Each sensing axis has its own dedicated surface map. For many devices and specifically in this spherical joint, the circular distribution of sensors

5.2 Field Mapping and Segmentation

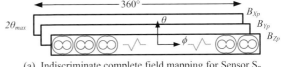

(a) Indiscriminate complete field mapping for Sensor S_p

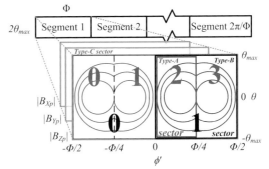

(b) Sectorization and segmentation for type-A,B and C symmetry

Fig. 5.2 Different approaches in multi-axis field mapping

and spatial periodicity Φ about the spin axis allows the surface map to be divided into $2\pi/\Phi$ segments as shown in Fig. 5.2b. Within each segment, the localized map can be demarcated further into smaller sectors by considering only the magnitude of **B** and using the direction of **B** for sector selection and identification as discussed in [1]. Three different degrees of symmetry (in decreasing symmetry) are considered and visually illustrated in Fig. 5.2b:

- Type-A symmetry: Assumes that the field contours residing in the sector defined by domain of $-\theta_{max} \leq \theta \leq \theta_{max}$ and $0 \leq \phi' \leq \Phi/4$ is unique and is related to the other three sectors through reflection and translation.
- Type-B symmetry: Assumes that the field contours are symmetric about the θ-axis or related by a translation of $\Phi/2$ in the ϕ'-axis. A type-B sector is defined by domain of $-\theta_{max} \leq \theta \leq \theta_{max}$ and $0 \leq \phi' \leq \Phi/2$.
- Type-C symmetry: No symmetry exists within a segment but like type-A and B, multiple segments are indistinguishable. The domain of a type-C sector coincides with the domain of a segment: $-\theta_{max} \leq \theta \leq \theta_{max}$ and $-\Phi/2 \leq \phi' \leq \Phi/2$. where ϕ' denotes the localized ϕ coordinate within a segment. A surface map exhibiting the most stringent Type-A symmetry will also possess both Type-B and C symmetry. However, the converse does not hold.

While the preceding field mapping focused on the multiple sensing axes of a single sensor, it is possible to extend the concept of multi-axis mapping to establish multi-sensor field mapping in a network of sensors as depicted in Fig. 5.3. In multi-sensor field mapping, field measurements from multiple sensors in a network can be utilized in a collaborative and complementary fashion. For example, a network of 3 single-axis sensors can produce three independent surface maps of the system just as a single three-axis sensor would generate. If segments are used,

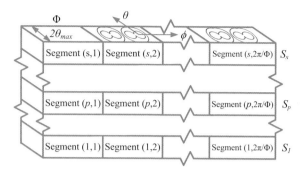

Fig. 5.3 Multi-sensor field mapping

segments from different sensors can be combined to create an overall surface map (segment 1 from sensor 1, segment 2 from sensor 3 and so on). In addition, by combining both approaches, an integrated approach of multi-axis and multi-sensor field mapping is also feasible (An arbitrary segment can be constructed using the X_1-axis surface map of sensor 1 and Y_5-axis surface map of sensor 5).

5.3 Artificial Neural Network Inverse Map

Two-layer neural networks utilizing the Levenberg-Marquardt supervised back-propagation algorithm are used to characterize the inverse mapping between the magnetic flux density measurements to the angular estimates of θ and ϕ. The entire surface map and sectionalized segments or sectors surface is discretized into a M × N grid, resulting to total of M × N training-target sets; 80% of the sets will be used for training, 15% for validation and 5% for testing. For this application, the h hidden nodes neural network has i inputs (magnetic field measurements) and j outputs (target position) and can be mathematically represented in (5.4):

$$[\hat{\psi}_v \quad \hat{\theta}_v \quad \hat{\phi}_v]^T = \mathbf{NN}_h(\{B_{X1}(v), B_{Y1}(v), \cdots, B_{Ys}(v), B_{Zs}(v)\}) \qquad (5.4)$$

where v is an integer representing the training set index ($1 \leq v \leq MN$), $i = 1, 2, \ldots, 3$, $j = 1, 2, 3$ and $\hat{\psi}_v$, $\hat{\theta}_v$, $\hat{\phi}_v$ are the angular estimates of the neural network. The mean squared error (MSE) is used to evaluate the performance of a neural network and is expressed as:

$$\mathrm{MSE} = \frac{1}{MN} \sum_{v=1}^{MN} \left[\left(\psi_v - \hat{\psi}_v\right)^2 + \left(\theta_v - \hat{\theta}_v\right)^2 + \left(\phi_v - \hat{\phi}_v\right)^2 \right] \qquad (5.5)$$

5.4 Experimental Investigation

Figure 5.4 shows a prototype spherical joint/actuator, where the two-layered 24—PM assembly embedded in the rotor are N52 Neodymium PMs. Due to the arrangement of these PMs, the spatial periodicity is 60° which corresponds to 6 segments (from Fig. 5.3). Underneath the PMs is a separate assembly of electromagnets (EMs) on the stator, which when energized is used for actuation. As the location and input current of each EM is known and an iron-free structure, superposition of multiple magnetic fields is valid and allows active compensation of the EM field in the sensor measurements during operation.

In order to perform sensor calibration on the multi-DOF spherical actuator, the rotor is rigidly attached to a rotary track of radius R by means of a mechanical strut as shown in Fig. 5.5. The center of rotary track is positioned such that it coincides with the spherical bearing of the rotor. The arc length of the track contains measurement markings that allow correspondence between the curvature distance, w and inclination, θ. A dual-axis MEMS inclinometer (VTI Tech. SCA121T) provides an independent measurement as a direct comparison and an optical incremental encoder (Kübler T8.A02H) affixed onto the strut measures the rotor spin motion.

Two types of magnetic sensors were used: a 3-Axis Hall-effect magnetic field sensor (Ametes MFS-3A) and a modified 2-axis Giant Magneto-resistance (GMR) (NVE Corp. AA003-02) which was prepared by attaching two single-axis GMR sensors. Table 5.1 summarizes the specifications of all sensors. The output of the magnetic sensors and inclinometer are transmitted in analog format as voltages digitally acquired using 16-bit (15-bit signed format) A/D converter banks.

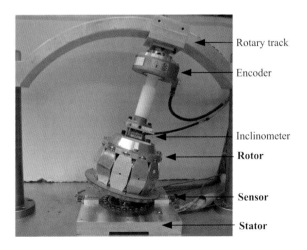

Fig. 5.4 Experimental setup

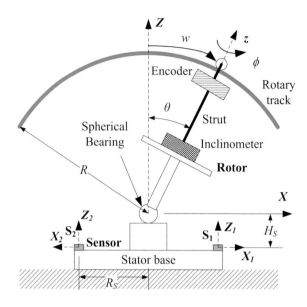

Fig. 5.5 Setup schematic

Table 5.1 Sensor specifications

	GMR	Hall	Inclinometer	Encoder
Type	Unipolar 2-axis (YZ)	Bipolar 3-axis (XYZ)	MEMS Dual-axis	Incremental quadrature
Range	1.4 mT	±7.3 mT	±30°	unlimited
Sensitivity	320 mV/mT	280 mV/mT	70 mV/°	4000 counts/rev
Bandwidth (−3 dB)	1 MHz	100 kHz	18 Hz	–
Price	$16	$24	$65 (chip)	–

A 2-D surface map depicting the field measurements at arbitrary inclination and spin of the rotor is achieved using both the rotary track and encoder. The rotor was preset to known inclinations (47 discrete points) using the rotary track and the field measurements **B** of the magnetic Hall sensors were recorded as the spin of the rotor is manually rotated at 0.36° increments (1000 data points per inclination set point). A complete surface map (2-D image with 47 × 1000 pixels) for the Y_p-axis measurement is shown in Fig. 5.6 along with the three sector types: Type A (47 × 42 pixels), Type B (47 × 84 pixels) and Type C (47 × 167 pixels). Six distinct segments can be easily detected in Fig. 5.6a and are numbered using Roman numerals. While sectors in B_Y can be classified up to Type-A symmetry, the

5.4 Experimental Investigation

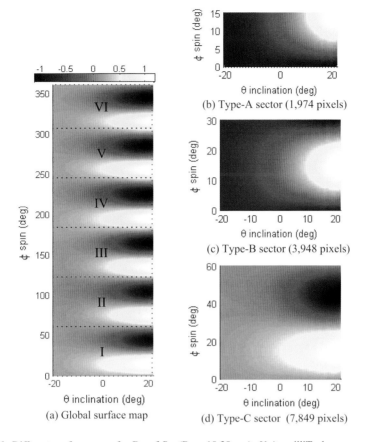

Fig. 5.6 Different surface maps for B_Y of S_1. ($R_s = 10.38$ cm). *Units* milliTesla

sectors in B_X and B_Z can only be categorized up to Type-B symmetry due to the offset placement (H_s) of the sensors.

Using segment I as the primary training segment, the Type-B and C sectors for each sensing axis of both sensors can be used to train ANNs with h = 50. The spatial distribution of the absolute estimation error from the trained ANN using all 6 axes from both sensors and presuming Type-B symmetry is shown in Fig. 5.7. Assuming the respective symmetry holds within segment I and replicated in other segments, the MSE resulting from operating an ANN trained for segment I using data in each of the remaining 5 segments are computed in Table 5.2. The MSE obtained from a trained ANN that assumes no spatial periodicity or symmetry in field measurements is provided for comparison. Also included in the table are the corresponding MSE if only B_X and B_Z (from each sensor) were utilized during training.

The results clearly suggest that ignoring the inherent symmetry of the system results in high MSE of the trained ANN. Assuming Type-B and C symmetry

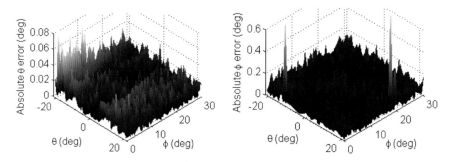

Fig. 5.7 Absolute spatial error distribution for inclination (top) and spin (bottom) estimates of segment I (h = 75, Type-B sector)

Table 5.2 Effects of symmetry on MSE (Segment I: primary)

	MSE (deg^2)						
h = 50	$S_{1,2}$ $B_{X,Z}$			$S_{1,2}$ $B_{X,Y,Z}$			
Symmetry	Type-B	Type-C	None	Type-B	Type-C	None	Ind. C
Sector	0.00525	0.954		0.00262	0.00826		
Seg. I	1.38			9.75			0.0082
Seg. II	2.32	702		11.3	36.8		0.0096
Seg. III	17.65	1140		15.0	51.1		0.0113
Seg. IV	5.41	1470		13.4	3.78		0.0091
Seg. V	3.95	865		7.64	39.2		0.0073
Seg. VI	9.98	3190		19.3	96.2		0.0101
Global	6.78	1230	2340	12.7	37.8	498	0.0092

reduces the overall MSE significantly in the sector/segment which was used for training while recycling the trained ANN for operation in other segments is some orders of magnitude higher due to variation in strength of magnetic fields in PMs. However, these errors are still much lower than an indiscriminate map of the entire global surface.

To achieve the lowest overall MSE, each segment (Type-C) should be individually trained using independent ANNs (in this case, 6 distinct ANNs), resulting in a more spatially consistent MSE as shown in the right-most column of Table 5.2. Hence during the operation all six ANNs will be used separately depending on the estimated spin position of the rotor obtained from a filter/predictor.

References

1. S. Foong, K.-M. Lee, Magnetic field-based multi-dof orientation sensor for PM-based spherical actuators. IEEE/ASME Adv. Intell. Mechatron., 481–486 (2009)
2. S. Foong, K.M. Lee, K. Bai, Magnetic field-based sensing method for spherical joint, in *Robotics and Automation (ICRA), 2010 IEEE International Conference on,* 2010, pp. 5447–5452

Chapter 6
A Back-EMF Method for Multi-DOF Motion Detection

This chapter presents a new method, referred to here as back-EMF method [1, 2], for multi-DOF motion sensing of a PMSM. With an explicit model characterizing the magnetic flux in the PMSM, a relationship between the electromotive force (EMF) induced in the winding of EMs and the motion of the rotor permanent magnets is derived; and closed-form solutions that solves for multi-DOF Euler angles and angular velocities of the rotor are presented. This method allows for simultaneous estimation of both quantities in real-time by only measuring the voltages across the EMs. Requiring no installation of additional sensors or fixtures on the rotor, the back-EMF method retains the structural simplicity of the PMSM.

6.1 Back-EMF for Multi-DOF Motion Sensing

Consider Fig. 6.1a where a closed-loop conductor C (with cross-sectional area of a) surrounded by a magnetic field (denoted with magnetic flux density **B**). The magnetic flux Φ through S (enclosed by C with surface normal **n**) is

$$\Phi = \int_S (\mathbf{B} \cdot \mathbf{n}) ds. \tag{6.1}$$

With the magnetic vector potential **A** defined in (6.2a) and the Stokes' theorem, the surface integral (6.1) can be reduced to a line integral (6.2b) where $\mathbf{l}$ is the directional vector of C:

$$\nabla \times \mathbf{A} = \mathbf{B}; \quad \Phi = \int_C (\mathbf{A} \cdot \mathbf{l}) dc \tag{6.2a, b}$$

© Huazhong University of Science and Technology Press, Wuhan and Springer Nature Singapore Pte Ltd. 2018
K. Bai and K.-M. Lee, *Permanent Magnet Spherical Motors*, Research on Intelligent Manufacturing, https://doi.org/10.1007/978-981-10-7962-7_6

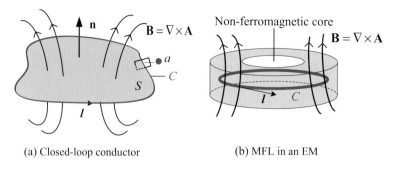

(a) Closed-loop conductor (b) MFL in an EM

Fig. 6.1 Illustration of magnetic flux computation

For an EM which can be treated as a contiguous filamentary conductor (Fig. 6.1b), the total magnetic flux linkage (MFL) through the EM winding can be obtained by extending (6.2b) to a volume integral (6.3) where V is the volume of the EM winding:

$$\Lambda = \frac{1}{a} \int_V (\mathbf{A} \cdot \mathbf{l}) dv \qquad (6.3)$$

For an electromagnetic motion system consisting of both PMs and EMs, the MFL in an EM is contributed by the magnetic fields from both the PMs and EMs in the system, therefore

$$\Lambda = \Lambda_P + \Lambda_E \text{ and } \mathbf{A} = \mathbf{A}_P + \mathbf{A}_E, \qquad (6.4a, b)$$

where

$$\Lambda_P = \frac{1}{a} \int_V (\mathbf{A}_P \cdot \mathbf{l}) dv \text{ and } \Lambda_E = \frac{1}{a} \int_V (\mathbf{A}_E \cdot \mathbf{l}) dv. \qquad (6.4c, d)$$

The equations for computing the magnetic vector potentials of a PM and an EM are given in (2.15) and (2.7).

The electromotive force (EMF) ε induced in an EM is contributed by two parts; ε_P due to the motion of the rotor PMs, and ε_E because of changes in current inputs (applied on the self- and mutual-inductances) of the EMs. According to the Faraday's Law of induction,

$$\varepsilon = -\frac{d\Lambda}{dt} = \varepsilon_P + \varepsilon_E, \text{ where } \varepsilon_P = -\frac{d\Lambda_P}{dt} \text{ and } \varepsilon_E = -\frac{d\Lambda_E}{dt}. \qquad (6.5a, b, c)$$

For systems with constant inductances, ε_E is known when the current inputs through the EMs are specified. This provides a basis for developing an explicit

6.1 Back-EMF for Multi-DOF Motion Sensing

model for a PMSM to characterize the relationship between its multi-DOF rotor motion and the EMFs in the stator EMs; and hence the rotor motion states can be acquired by measuring the voltages in the EMs.

6.1.1 EMF Model in a Single EM-PM Pair

Figure 6.2 shows an EM-PM pair in a PMSM, where both PM and EM are cylindrical; *XYZ* and *xyz* (sharing a common origin) are the stator reference and rotor local frames respectively; **s** and **r** represent the position vectors in *XYZ* frame from the origin to the geometrical centers of the stator EM and rotor PM; and σ is the separation angle between **s** and **r**. In Fig. 6.2, y' and x'' are two intermediate axes for defining *XYZ* Euler angles (α, β, γ). Given the PM position vector **p** in *xyz* coordinates, **r** can be obtained from the rotation matrix [**R**] in (6.6a, b) where S and C represent sine and cosine of the Euler angles (subscripts) respectively:

$$\mathbf{r} = [\mathbf{R}]^T \mathbf{p}, \tag{6.6a}$$

where

$$[\mathbf{R}] = \begin{bmatrix} C_\gamma C_\beta & S_\gamma C_\alpha + C_\gamma S_\beta S_\alpha & S_\gamma S_\alpha - C_\gamma S_\beta C_\alpha \\ -S_\gamma C_\beta & C_\gamma C_\alpha - S_\gamma S_\beta S_\alpha & C_\gamma S_\alpha + S_\gamma S_\beta C_\alpha \\ S_\beta & -C_\beta S_\alpha & C_\beta C_\alpha \end{bmatrix}. \tag{6.6b}$$

As shown in Fig. 6.2 the MFL contributed by the PM in the EM winding is only a function f of the separation angle σ, which can be derived from (6.4c) by normalizing with respect to the sign of the PM magnetization λ:

$$f(\sigma) = \frac{\Lambda_P}{\lambda} = \frac{1}{a\lambda} \int_V (\mathbf{A}_P \cdot \mathbf{l}) dv \text{ where } \sigma = \cos^{-1}(\mathbf{e}_r \cdot \mathbf{e}_s), \quad \mathbf{e}_r = \frac{\mathbf{r}}{|\mathbf{r}|} \text{ and } \mathbf{e}_s = \frac{\mathbf{s}}{|\mathbf{s}|}.$$

$$\tag{6.7a, b, c, d}$$

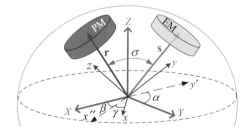

Fig. 6.2 Single-EM/PM system

Similarly, the self-inductance L of the EM with current input I can be derived from (6.4d) as

$$L = \frac{\Lambda_E}{I} = \frac{1}{aI}\int_V (\mathbf{A}_E \cdot \mathbf{l})dv. \tag{6.8}$$

Since the voltage U across the current-carrying EM (with resistance R) can be represented with (6.9) where σ is from (6.7b), the induced EMF ε can be derived as (6.10) using chain rule by substituting Λ_P and Λ_E from (6.7a) and (6.8) respectively into (6.5a):

$$U = RI - \varepsilon \quad \text{where } \varepsilon = -\frac{d[\lambda f(\sigma)]}{dt} - \frac{d(LI)}{dt}; \tag{6.9}$$

$$\varepsilon = \frac{\lambda f'(\sigma)}{\sqrt{1 - (\mathbf{e}_r \cdot \mathbf{e}_s)^2}}(\dot{\mathbf{e}}_r \cdot \mathbf{e}_s) - L\dot{I}, \quad \text{where } f'(\sigma) = df/d\sigma. \tag{6.10}$$

As the current I is a specified (and thus known) input to the electromagnetic system, the PM position $\mathbf{e}_r$ and its derivative $\dot{\mathbf{e}}_r$ can be deduced by measuring U (in a back-EMF sense). Furthermore, if f and its derivative are explicit functions of σ for a PMSM, $\mathbf{e}_r$ and $\dot{\mathbf{e}}_r$ can be derived and expressed in closed form which will greatly facilitate the real time implementation of the back-EMF method.

6.1.2 Back-EMF with Multiple EM-PM Pairs

For a PMSM consisting of N_P PMs and N_E EMs, the MFL of each pole-pair (Fig. 6.2) can be characterized by the relationships (6.7–6.10). The total MFL in the jth EM is contributed by all the EMs and PMs:

$$\Lambda_j = \sum_{l=1}^{N_P} \lambda_l f(\sigma_{j,l}) + \sum_{m=1}^{N_E} L_{j,m} I_m. \tag{6.11}$$

To simplify notations, the subscripts, j and l, refer to jth EM (EM$_j$) and lth PM (PM$_l$) respectively. In (6.11), the separation angle $\sigma_{j,l}$ between EM$_j$ and PM$_l$ can be computed using (6.7b); $L_{j,m}$ is the mutual-inductance between EM$_j$ and EM$_m$ (when $j \neq m$) or self-inductance of EM$_j$ (when $j = m$), which can be computed using (6.19) in the appendix. Along with the equations for the rotation matrix (6.6b), the total induced voltage in EM$_j$ can be derived in terms of the rotor orientation vector $\mathbf{q} = [\alpha\ \beta\ \gamma]^T$ in a similar manner as (6.10):

6.1 Back-EMF for Multi-DOF Motion Sensing

$$\varepsilon_j = \mathbf{V}_j^T \dot{\mathbf{q}} - \sum_{m=1}^{N_E} L_{j,m} \dot{I}_m \quad \text{where } \mathbf{V}_j = \begin{bmatrix} \sum_{l=1}^{N_P} \Theta_{lj} (\mathbf{a}\mathbf{e}_{pl})^T \mathbf{e}_{sj} \\ \sum_{l=1}^{N_P} \Theta_{lj} (\mathbf{b}\mathbf{e}_{pl})^T \mathbf{e}_{sj} \\ \sum_{l=1}^{N_P} \Theta_{lj} (\mathbf{c}\mathbf{e}_{pl})^T \mathbf{e}_{sj} \end{bmatrix}; \quad (6.12\text{a},\text{b},\text{c})$$

$$\text{and } \Theta_{lj} = \frac{\lambda_l f'(\sigma_{lj})}{\sqrt{1-(\mathbf{e}_{rl} \cdot \mathbf{e}_{sj})^2}}.$$

In (6.12b), $\mathbf{e}_{pl}$ (constant) and $\mathbf{e}_{rl}$ (varying) are the unit vectors pointing PM_l in xyz and XYZ frames respectively; $\mathbf{e}_{sj}$ (constant) is the unit vector pointing EM_j; **a**, **b**, **c** are given as follows:

$$\mathbf{a} = \begin{bmatrix} 0 & 0 & 0 \\ -S_\gamma S_\alpha + C_\gamma S_\beta C_\alpha & -(C_\gamma S_\alpha + S_\gamma S_\beta C_\alpha) & -C_\beta C_\alpha \\ S_\gamma C_\alpha + C_\gamma S_\beta S_\alpha & C_\gamma C_\alpha - S_\gamma S_\beta S_\alpha & -C_\beta S_\alpha \end{bmatrix};$$

$$\mathbf{b} = \begin{bmatrix} -C_\gamma S_\beta & S_\gamma S_\beta & C_\beta \\ C_\gamma C_\beta S_\alpha & -S_\gamma C_\beta S_\alpha & S_\beta S_\alpha \\ -C_\gamma C_\beta C_\alpha & S_\gamma C_\beta C_\alpha & -S_\beta C_\alpha \end{bmatrix};$$

$$\mathbf{c} = \begin{bmatrix} -S_\gamma C_\beta & -C_\gamma C_\beta & 0 \\ C_\gamma C_\alpha - S_\gamma S_\beta S_\alpha & -(S_\gamma C_\alpha + C_\gamma S_\beta S_\alpha) & 0 \\ C_\gamma S_\alpha + S_\gamma S_\beta C_\alpha & -S_\gamma S_\alpha + C_\gamma S_\beta C_\alpha & 0 \end{bmatrix}.$$

Therefore, the total voltage across EM_j is

$$U_j = R_j I_j + \sum_{m=1}^{N_E} L_{j,m} \dot{I}_m - \mathbf{V}_j^T \dot{\mathbf{q}}. \quad (6.13)$$

Accounting for multiple EMs, the back-EMF model solving for the angular velocity of the Euler angles from measured voltages across the EMs is given by (6.14) where [**W**] depends on **q**:

$$[\mathbf{W}]\dot{\mathbf{q}} = \mathbf{Y}, \quad (6.14)$$

where

$$\mathbf{Y} = \begin{bmatrix} \vdots \\ -U_j + I_j R_j + \sum_{m=1}^{N_E} L_{j,m} \dot{I}_m \\ \vdots \end{bmatrix} \text{ and } [\mathbf{W}] = \begin{bmatrix} \vdots \\ \mathbf{V}_j^{\mathrm{T}} \\ \vdots \end{bmatrix}. \qquad (6.14\text{a, b})$$

For a discrete-time system, the solution to (6.14) at kth time instant can be solved using a least-square method that leads to a pseudo-inverse solution (6.15):

$$\dot{\mathbf{q}}_k = ([\mathbf{W}_k]^T [\mathbf{W}_k])^{-1} [\mathbf{W}_k]^T \mathbf{Y}_k. \qquad (6.15)$$

Therefore, the orientation and its angular velocities can be acquired incrementally (with sampling time Δt) for a given initial orientation $\mathbf{q}_0$:

$$\mathbf{q}_k = \begin{cases} \mathbf{q}_0 & k = 0 \\ \mathbf{q}_{k-1} + \dot{\mathbf{q}}_{k-1} \Delta t & k = 1, 2, \ldots \end{cases} \qquad (6.16)$$

6.2 Implementation of Back-EMF Method on a PMSM

The back-EMF method has been implemented on a PMSM as shown in Fig. 6.3. The numerical models and their solutions for implementation are presented and the feasibility of the back-EMF method using measured EM voltages for simultaneous estimation of both orientation and angular velocity are experimentally verified. The estimation results are compared with a commercialized gyroscope. As an immediate application, the motion states acquired with the back-EMF method are used for parameter estimation of the PMSM.

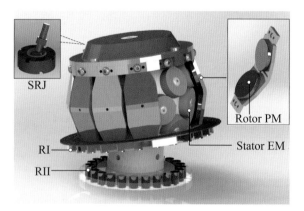

Fig. 6.3 CAD model of PMSM

6.2 Implementation of Back-EMF Method on a PMSM

6.2.1 Mechanical and Magnetic Structure of the PMSM

The ball-joint-like PMSM as shown in Fig. 6.3 consists of a spherical stator (inside) and a socket-like rotor (outside). The stator consists of three layers of 8 cylindrical EMs whereas the rotor has two layers of 12 cylindrical PMs. In spherical coordinates, the centroids of the EMs and PMs are defined in stator XYZ and rotor xyz frames respectively in (6.17):

$$\vec{C} = [R \sin \Theta \cos \Phi, \quad R \sin \Theta \sin \Phi, \quad R \cos \Theta]^T \tag{6.17}$$

In (6.17), R, Θ, Φ are the radial distance, azimuth/polar angles respectively. The parameters describing the PMs and EMs are shown in Table 6.1.

Table 6.1 PMSM parameters

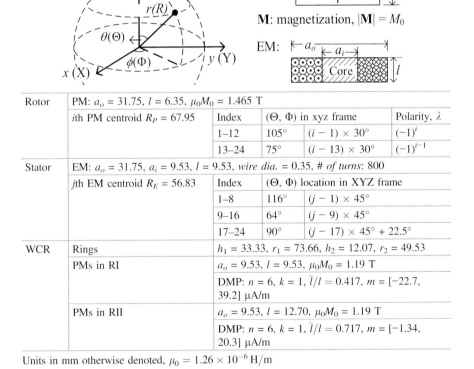

Rotor	PM: $a_o = 31.75$, $l = 6.35$, $\mu_0 M_0 = 1.465$ T				
	ith PM centroid $R_P = 67.95$	Index	(Θ, Φ) in xyz frame		Polarity, λ
		1–12	105°	$(i - 1) \times 30°$	$(-1)^i$
		13–24	75°	$(i - 13) \times 30°$	$(-1)^{i-1}$
Stator	EM: $a_o = 31.75$, $a_i = 9.53$, $l = 9.53$, wire dia. $= 0.35$, # of turns: 800				
	jth EM centroid $R_E = 56.83$	Index	(Θ, Φ) location in XYZ frame		
		1–8	116°	$(j - 1) \times 45°$	
		9–16	64°	$(j - 9) \times 45°$	
		17–24	90°	$(j - 17) \times 45° + 22.5°$	
WCR	Rings	$h_1 = 33.33$, $r_1 = 73.66$, $h_2 = 12.07$, $r_2 = 49.53$			
	PMs in RI	$a_o = 9.53$, $l = 9.53$, $\mu_0 M_0 = 1.19$ T			
		DMP: $n = 6$, $k = 1$, $\bar{l}/l = 0.417$, $m = [-22.7, 39.2]$ μA/m			
	PMs in RII	$a_o = 9.53$, $l = 12.70$, $\mu_0 M_0 = 1.19$ T			
		DMP: $n = 6$, $k = 1$, $\bar{l}/l = 0.717$, $m = [-1.34, 20.3]$ μA/m			

Units in mm otherwise denoted, $\mu_0 = 1.26 \times 10^{-6}$ H/m

Fig. 6.4 Restoring torque of WCR

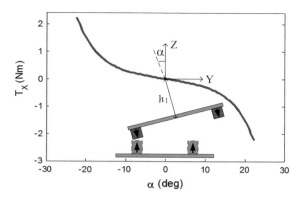

The weight compensating regulator (WCR), as presented in Sect. 3.5 (in Chap. 3), is incorporated for the PMSM to angularly support its rotor against gravity by means of magnetic repulsion using two circular rings of PMs indicated as RI and RII in Fig. 6.3. The detailed description of the WCR as well as the procedures for computing the restoring torque can be found in Sect. 3.5; and the restoring torque is plotted in Fig. 6.4.

6.2.2 Numerical Solutions for the MFL Model

For the implementation of the back-EMF method, the numerical models as well as their solutions are provided.

Numerical Model of MFL in an EM-PM Pair
Figure 6.5 shows the computed results of the MFL model (6.7a) as a function of the separation angle σ for the single EM/PM pair (Fig. 6.2) of the PMSM, where the parametric values are listed in Table 6.1. For verification, the results are compared against those obtained using *ANSYS* in Fig. 6.5, which agree very well. In order to derive a closed-form expression of the induced EMF for real-time computation, $f(\sigma)$ is curve-fitted with a Fourier-form function by matching the exact solution computed using numerical integration; the resulting fits for $f(\sigma)$ and its derivative are given in (6.18a,b) and plotted in Fig. 6.5:

$$f(\sigma) = \sum_{j=0}^{7} \left[a_j \sin(jw\sigma) + b_j \cos(jw\sigma) \right]$$

$$\text{and } f'(\sigma) = \sum_{j=1}^{7} jw \left[a_j \cos(jw\sigma) - b_j \sin(jw\sigma) \right].$$

(6.18a, b)

6.2 Implementation of Back-EMF Method on a PMSM

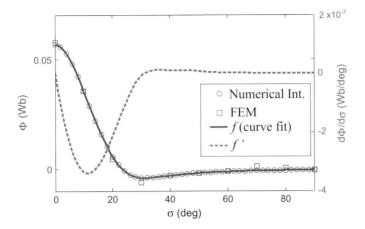

Fig. 6.5 Flux linkage model and curve-fit approximation

where

$$w = 0.05;$$
$$[a_1, \ldots, a_7] = [-0.18, 1.52, 1.24, 0.09, -0.51, -0.45, -0.17] \times 10^{-3};$$
$$[b_0, \ldots, b_7] = [9.5, 19.8, 15.0, 8.3, 3.3, 0.85, 0.06, -0.02] \times 10^{-3}$$

Computation of Self and Mutual Inductances of the EMs

The self- and mutual-inductances can be computed using (6.8) and (A.1) in the appendix. As the inductances only depend on the relative positions of the EMs, the self- and mutual-inductances of the EMs in the PMSM stator are computed as a function of the separation angle, which is shown in Fig. 6.6a. Note that the self-inductances correspond to EMs with separation angle at zero. For illustration, Fig. 6.6b depicts the EM locations and indexes. The 24 EMs in the stator totally result in 576 (24 × 24) self/mutual-inductances. Due to symmetry, the inductances

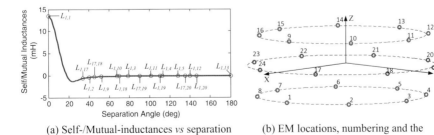

(a) Self-/Mutual-inductances *vs* separation angle

(b) EM locations, numbering and the separation angles

Fig. 6.6 Self-/mutual-inductances of the EMs

lead to 17 different values which are denoted in Fig. 6.6a. For illustration, inductances involving EM_1 and EM_{17} are labeled on the figure.

6.2.3 Experiment and Discussion

The back-EMF method was experimentally implemented on a prototype PMSM as shown in Fig. 6.7. The algorithms involved in the back-EMF method are implemented on an *NI CRIO-9025* real-time unit featured with 800 MHz processor and 512 RAM where the EM voltages in the PMSM are acquired with the A/D modules in real-time.

The procedure for real-time implementation of the back-EMF method can be summarized with the following steps where Step (0) is performed offline:

(0) Offline: Compute and normalize the MFL with (6.7a) and derive a closed-form expression for function f and its derivative f' (in the form of Eqs. 6.18a,b); compute the self-/mutual-inductances L's of the EMs in the stator with (A.1) in the appendix;

(1) Get the initial/updated orientation. For the first iteration, use initial orientation $\mathbf{q}_0$ as the present orientation; for the rest iterations, use the updated orientation from previous iteration.

(2) Form $\mathbf{Y}$ (vector) following (6.14a) with measured EM voltages and the current inputs as well as the computed inductances; Form $[\mathbf{W}]$ (matrix) following (6.14b) where the component $\mathbf{V}_j$ is computed using (6.12b) with present Euler angles and function f';

(3) Use (6.15) to compute the angular velocity with $\mathbf{Y}$ and $[\mathbf{W}]$;

(4) Update the orientation using (6.16) with the computed angular velocity.

In this experiment, the rotor was given an initial orientation and then released. It oscillated about its equilibrium due to the restoring torque of the WCR. During the

Fig. 6.7 PMSM test-bed (gyroscope was for verification only)

6.2 Implementation of Back-EMF Method on a PMSM

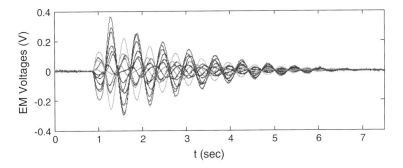

Fig. 6.8 Measured EM voltages

process, EM voltages were recorded (plotted in Fig. 6.8) and the Euler angles and angular velocities were estimated using the back-EMF method (6.15–6.16). For verification, the estimated results were compared with measurements from a commercialized gyroscope (*LYPR540AH*) mounted on the rotor.

In order to investigate the effects of the total number of EMs used for back-EMF on estimation accuracy, two design configurations were considered where the locations and numbering of the EMs are shown in Fig. 6.6b:

DC1: All 24 EMs are used for back-EMF sensing;
DC2: Only 8 EMs in the mid-layer (EMs 17-24) are used for back-EMF sensing.

Note that DC2 also offers an alternative means for the PMSM operation where the mid-layer EMs can be used as dedicated sensing coils and only the upper and

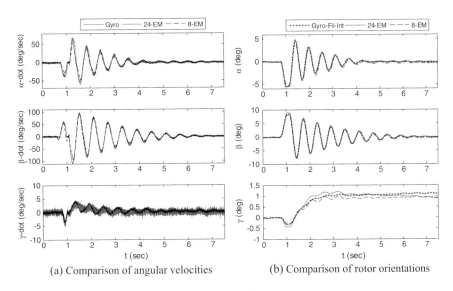

(a) Comparison of angular velocities   (b) Comparison of rotor orientations

Fig. 6.9 Comparison between back-EMF method and a gyroscope

Table 6.2 Percentage errors of Euler angles and angular velocities

Configurations	α	β	γ	$d\alpha/dt$	$d\beta/dt$	$d\gamma/dt$
24-EM	2.36	1.94	3.87	3.78	2.58	3.06
8-EM	5.82	4.08	5.21	4.44	3.71	5.02

bottom layer EMs are used for actuation. In this case, (6.13) can be reduced by letting the current inputs equal to zero, which will greatly reduce the computational and hardware costs in implementation.

Figure 6.9a compares the angular velocities estimated using the back-EMF against those acquired from the gyroscope. In order to compare the orientations, the angular velocities acquired from the gyroscope were filtered and integrated to obtain the Euler angles. These results are compared with the rotor orientation estimated using the back-EMF method in Fig. 6.9b. In order to evaluate the accuracy of the back-EMF method, the relative mean absolute error (*RMAE*) were calculated for the estimated Euler angles and angular velocities shown in Fig. 6.9a, b; and the results are summarized in Table 6.2.

Some observations of Fig. 6.9a, b as well as Table 6.2 can be drawn as follows:

- Both angular velocities and Euler angles estimated with the back-EMF method match the results acquired from the gyroscope very well where the maximum percentage errors are less than 6%.
- DC1 using more EMs offers better accuracy; DC2 using only 8 EMs also shows comparable estimation results and hence provides a cost-effective sensing solution.
- The angular velocities estimated with back-EMF show smaller noise-to-signal ratio than the results acquired with the gyroscope in both cases.

6.2.4 Parameter Estimation of the PMSM with Back-EMF Method

As an immediate application, the back-EMF method is used for estimating the uncertain parameters (mass/inertia properties and damping) of the PMSM dynamic model for its capability of capturing all motion states without installing any external sensors and fixtures on the rotor; these installations introduce unknown inertia and lead to inaccurate estimation results.

For parameter estimation, the gyroscope was taken off from the rotor and a natural response was initiated with the rotor released from an initial orientation. The restoring torque (from the WCR) was computed at the acquired orientations following SI and SII in Appendix. Given the dynamic model of the PMSM (in Sect. 2.2), the acquired motion states and restoring torques were loaded to the *Matlab System Identification Toolbox (SIT)* to estimate the uncertain parameters.

6.2 Implementation of Back-EMF Method on a PMSM

Table 6.3 Parameter estimation results

	I_a (kg m^2)	I_t (kg m^2)	b (kg/s)	$m_r g l_r$ (N m)
Ini. guesses	8.23×10^{-3}	6.26×10^{-3}	0	0.064
Est. values	1.23×10^{-2}	7.9×10^{-3}	0.011	0.18

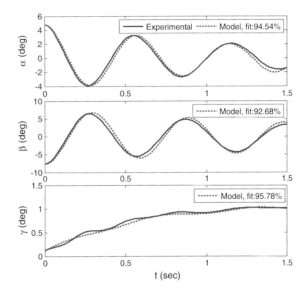

Fig. 6.10 Parameter estimation results

Note that the frictional torque of the bearing is approximated with a pure damping (with coefficient b). The mass properties of the CAD model of the PMSM obtained using *Solidworks* were used for initial guesses for the *SIT* and also provide a reference for comparison. Table 6.3 lists the initial guesses and estimated values of the target parameters. With the estimated parameters, the dynamic response of the rotor orientation was simulated which was compared with the measured rotor orientation in Fig. 6.10. The fit-values denoted in Fig. 6.10 provided by the *SIT* which can be considered a criterion of the estimation performance suggests that the dynamic model with the estimated parameters match with the real system closely.

Appendix

Computation of Self-/Mutual Inductances
The self-/mutual inductances can be computed as

$$L_{j,m} = \frac{\Lambda_E}{I_m} = \frac{1}{aI_m} \int_V (\mathbf{A}_{Ej} \cdot \mathbf{l}) dv \qquad (6.19)$$

where $L_{j,m}$ is the mutual-inductance between EM_j and EM_m (when $j \neq m$) or self-inductance of EM_j (when $j = m$); a is the cross-sectional area of winding wire; I_m is the current in EM_m; $\mathbf{A}_{Ej}$ is the magnetic vector potential generated by EM_j.

References

1. K. Bai, K.-M. Lee, J. Lu, A magnetic flux model based method for detecting multi-DOF motion of a permanent magnet spherical motor. Mechatronics **39**, 217–225 (2016)
2. K. Bai, K.-M. Lee, A sensor-less motion sensing method of a 3-DOF permanent magnet spherical motor, in *IFAC Proceedings Volumes, 6th IFAC Symposium on Mechatronic Systems*, vol. 46, no. 5, pp. 160–164, Hangzhou, China, 10–12 April 2013

Part III
Control Methods

Chapter 7
Direct Field-Feedback Control

With the rotor dynamics of PMSMs presented in Chap. 2 (Sect. 2.2), the conventional control methods for controlling the multi-DOF orientation of a PMSM are introduced at the beginning of this chapter. The practical challenges in implementing the conventional control systems for PMSMs are illustrated and an alternative method, the direct field-feedback control (DFC) method, is presented. The major components of this new method are described and compared to conventional control systems for a PMSM. The theoretical framework and the implementing procedures of the proposed method are presented and illustrated in details with both numerical simulations and experiments.

7.1 Traditional Orientation Control Method for Spherical Motors

Figure 7.1 presents a conventional orientation-dependent control system for a PMSM. In Fig. 7.1, the control law determines the desired torque $\mathbf{T}_d$ in order to track an orientation input $\mathbf{q}_d$ based on the error $\mathbf{e_q}$ between the desired and measured orientations. In order to determine the current input to generate the desired torque $\mathbf{T}_d$, the torque-to-current relationship is required. Based on the forward torque model presented in (4.22) in Sect. 4.3, the torque characteristic vectors (TCVs included in the matrix $[\mathbf{K}]$) are computed through an orientation-dependent model, as shown in Fig. 7.1. The optimal current vector $\mathbf{u}$ to realize the desired torque can be found using (4.23).

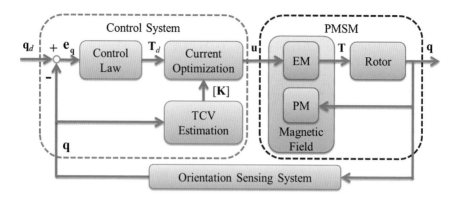

Fig. 7.1 Orientation-feedback control system

7.1.1 PD Control Law and Stability Analysis

For a multi-DOF rotational system, most control laws that can be applied are based on or have a similar form to a PD control. A proportional-derivative (PD) control law that can be used to regulate the PMSM rotor at an equilibrium takes the form:

$$\mathbf{T} = \mathbf{K}'_P \mathbf{e_q} + \mathbf{K}'_D \dot{\mathbf{e}}_\mathbf{q} \tag{7.1}$$

For the rotor dynamic (2.37) and the PD control law, let the Lyapunov function be the virtual mechanical energy with the form:

$$V = \frac{1}{2}\left(\dot{\mathbf{q}}^T[\mathbf{M}]\dot{\mathbf{q}} + \mathbf{e}_\mathbf{q}^T[\mathbf{K}'_P]\mathbf{e_q}\right) \tag{7.2}$$

Meanwhile, the conservation of energy can be written as

$$\frac{1}{2}\frac{d}{dt}\left(\dot{\mathbf{q}}^T[\mathbf{M}]\dot{\mathbf{q}}\right) = \dot{\mathbf{q}}^T \cdot \mathbf{T} \tag{7.3}$$

where the left hand side is the derivative of the kinetic energy; and the right-hand side represents the power input of the motor. Since the inertia matrix [**M**] (illustrated in Eqs. 2.37a and 2.38a in Sect. 2.2) is symmetric positive definite, the time derivative of the Lyapunov function can be obtained with (7.3) as:

$$\dot{V} = \frac{1}{2}\frac{d}{dt}\left(\dot{\mathbf{q}}^T[\mathbf{M}]\dot{\mathbf{q}}\right) + \frac{1}{2}\frac{d}{dt}\left(\mathbf{e}_\mathbf{q}^T[\mathbf{K}'_P]\mathbf{e_q}\right) \tag{7.4}$$

Since $\dot{\mathbf{q}}_d = 0$ at the equilibrium state, substituting (7.1) and (7.3) into (7.4), yields:

7.1 Traditional Orientation Control Method for Spherical Motors

$$\dot{V} = -\dot{\mathbf{q}}^T [\mathbf{K}'_D] \dot{\mathbf{q}} \qquad (7.5)$$

Therefore, as long as the control matrices $\mathbf{K}'_P$ and $\mathbf{K}'_D$ in (7.1) are positive definite, $V > 0$, and $\dot{V} \leq 0$. Meanwhile, since $\dot{V} = 0$ implies that $\dot{\mathbf{q}} = 0$; along with (2.37), (7.1) and (7.3), $\dot{V}$ is identically 0 only if $\mathbf{e_q} = 0$. Therefore, the system is stable and converges to the desired state.

7.1.2 Comments on Implementation of Traditional Control Methods

The real-time control of a PMSM requires the computation of [**K**] which depends on the magnetic flux density (MFD) **B(q)** to determine the current vector for delivering the required torque. Traditionally, the PMSM controlled system relies on explicit orientation measurements to estimate **B(q)** and [**K**], and to derive an error $\mathbf{e_q}$ between the desired and measured orientation as shown in Fig. 7.1. Computational delay and cumulated errors as a result of serial estimation of **q**,**B(q)** and [**K**] significantly limit the dynamic performance of a PMSM controlled system.

The TCV computation in conventional control systems relies on orientation-based models, which can only be computationally processed after the orientation is explicitly determined. This sequential processing often downgrades the control performance because of accumulated errors and time delay in computations. There is a need to develop an efficient way to acquire the required control effort to improve the overall performance of a PMSM.

For the above reasons, the feasibility of utilizing real-time measurements and direct feedback of existing rotor magnetic field for multi-DOF orientation control is explored here. As a common media for energy conversion between electrical and mechanical systems, the magnetic field is invariant to environmental factors and can be acquired almost instantaneously with low-cost precision sensors that have a very small footprint. In a PMSM, the magnetic field of the rotor PMs is a function of rotor orientation, and has a direct relationship with the TCVs. Thus, the orientation feedback can be equivalently replaced with magnetic field feedback in control law and the TCVs can be estimated directly using measured magnetic fields, eliminating the need for an explicit orientation sensor. In addition, the control law derivation and TCV estimation only require magnetic field measurements, permitting simultaneous computation of both quantities. Such a control system featured with direct magnetic field feedback and parallel computation can greatly improve the overall control performance of a PMSM.

7.2 Direct Field-Feedback Control

Figure 7.2 shows the design concept of a direct field-feedback control (DFC) system [1] utilizing the existing MFD **B** of the rotor PMs for controlling the orientation **q** of a PMSM and thus eliminating the need for an external orientation sensor. In Fig. 7.2, **S** represents a specified set of magnetic sensors installed on the stator; and $\mathbf{B_S}$ is an MFD vector measured at the rotor orientation **q** by sensors in **S**. $\mathbf{B_S}$ and **q** have the same dimension as the mechanical DOF N_D of the system.

The DFC system consists of the following functional blocks; a forward **B**-model, **B**-based control law, field-based sensing system and **B**-based TCV estimation, and an inverse torque model. For a desired orientation $\mathbf{q}_d$, the forward **B**-model (that can be calibrated off-line) determines the corresponding MFD vector $\mathbf{B}_{Sd}$ at **S**. The control law utilizes the error signal $\mathbf{e_B}$ (between $\mathbf{B}_{Sd}$ and $\mathbf{B_S}$) to determine the desired torque $\mathbf{T}_d$ upon which an optimal current input vector **u** can be computed from the inverse torque model along with the measured MFDs and hence the estimated [**K**]. Both the control law and TCV estimation only require the measured MFD as shown in Fig. 7.2, and can be computed in parallel which greatly improve the real-time computational efficiency.

As will be illustrated experimentally, the MFDs of the EMs acquired by the magnetic sensors can be subtracted off from the MFD measurements (of the embedded sensors) in real-time. For simplicity, the $\mathbf{B_S}$ and $\mathbf{B}_{Sd}$ in a DFC system specifically refer to the rotor MFD in the following discussions.

To ensure that the rotor approaches the desired orientation $\mathbf{q}_d$ (that uniquely corresponds to $\mathbf{B}_{Sd}$) when the control system drives $\mathbf{B_S}$ to $\mathbf{B}_{Sd}$, the DFC must be designed such that **q** and $\mathbf{B_S}$ are bijective (one-to-one and onto). Due to the periodicity of the rotor PM placement (commonly found in a PMSM), the bijective domain for an individual sensor set is generally limited to a portion of the workspace. A multi-sensor network with connected bijective domains of multiple sensors to cover the entire workspace is often required. For developing a realizable DFC PMSM system, the following three key issues are discussed; bijective domain, control parameters and multi-sensor approach.

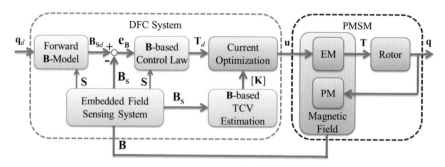

Fig. 7.2 Direct field-feedback control (DFC) system [1]

7.2 Direct Field-Feedback Control

7.2.1 Determination of Bijective Domain

Without loss of generality, $\mathbf{f}$ is defined as a function that matches $\mathbf{q}$ to $\mathbf{B_S}$:

$$\mathbf{B_S} = \mathbf{f}(\mathbf{q}) \tag{7.6}$$

According to the implicit function theorem [2], for $\mathbf{q}, \mathbf{B_S} \in \mathbb{R}^{N_D}$, $\mathbf{f}$ is bijective (and invertible) in a neighborhood around $\mathbf{q}_0$ if

$$J = \det([\mathbf{J}])|_{\mathbf{q}=\mathbf{q}_0} \neq 0 \tag{7.7}$$

In (7.7), the elements of the Jacobian matrix $\mathbf{J}$ have the form:

$$\mathbf{J} = [\partial B_i / \partial q_j] \quad \text{where} \quad i,j = 1, 2, \ldots, N_D \tag{7.8}$$

and B_i and q_j are the ith and jth components in $\mathbf{B_S}$ and $\mathbf{q}$ respectively. As explicit forms for $\mathbf{f}$ and $\mathbf{J}$ cannot be easily found, the bijective domain is determined numerically. The bijective condition (7.7) requires that the Jacobian matrix is nonsingular. For numerical elimination of singular Jacobian matrixes, two alternative conditions (based on the determinant and condition number) can be used to avoid incorrect determination due to errors in numerical approximations:

$$\text{Condition 1:} \quad \Gamma = \{\mathbf{q} | |J| > \varepsilon\} \tag{7.9a}$$

$$\text{Condition 2:} \quad \Gamma = \{\mathbf{q} | \kappa(\mathbf{J}) < \chi\}, \quad \text{where} \quad \kappa(\mathbf{J}) = \sigma_{\max}/\sigma_{\min} \tag{7.9b}$$

In (7.9), ε and χ are positive constants; $\kappa(\mathbf{J})$ is the condition number; and $\sigma_{\max}$ and $\sigma_{\min}$ are the maximum and minimum singular values of the Jacobian matrix.

7.2.2 DFC Control Law and Control Parameter Determination

For controlling the 3-DOF rotor dynamics, the PD control law of the DFC with MFD feedback takes the form:

$$\mathbf{T}_d = [\mathbf{K}_P]\mathbf{e_B} + [\mathbf{K}_D]\dot{\mathbf{e}}_B \quad \text{where} \quad \mathbf{e_B} = \mathbf{B}_{Sd} - \mathbf{B_S}. \tag{7.10}$$

In (7.10), $\mathbf{K}_P = \mathrm{diag}(k_{p1}, k_{p2}, k_{p3})$, and $\mathbf{K}_D = \mathrm{diag}(k_{d1}, k_{d2}, k_{d3})$. In a bijective domain, the orientation error $\mathbf{e_q}$ can be characterized in terms of $\mathbf{e_B}$:

$$\mathbf{e_B} = [\mathbf{A}]\mathbf{e_q} \quad \text{where} \quad [\mathbf{A}] = [a_{ij}] \quad \text{and} \quad \mathbf{e_q} = \mathbf{q}_d - \mathbf{q}. \tag{7.11}$$

In (7.11), a_{ij} is bounded by $\partial B_i/\partial q_j$ in each bijective domain [3]. Substituting (7.11) into (7.10), an auxiliary control law is obtained:

$$\mathbf{T}_d = [\mathbf{K}'_P]\mathbf{e_q} + [\mathbf{K}'_D]\dot{\mathbf{e}}_\mathbf{q} \tag{7.12}$$

For the dynamic system given in (2.38), as long as $[\mathbf{K}'_P] = [\mathbf{K}_P][\mathbf{A}]$ and $[\mathbf{K}'_D] = [\mathbf{K}_D][\mathbf{A}]$ are positive definite, the system is stable and its states will converge to the desired orientation. Consider $\mathbf{x} = \begin{bmatrix} x_1 & x_2 & x_3 \end{bmatrix}^T \neq 0$,

$$\mathbf{x}^T([\mathbf{K}_P][\mathbf{A}])\mathbf{x} = b_1 x_1^2 + b_2 x_2^2 + b_3 x_3^2 + b_{12} x_1 x_2 + b_{13} x_1 x_3 + b_{23} x_2 x_3 \tag{7.13}$$

where $b_i = k_{pi} a_{ii}$ and $b_{ij} = k_{pi} a_{ij} + k_{pj} a_{ji}$ $(i, j = 1, 2, 3)$.

For $b_{12}, b_{13}, b_{23} > 0 (\neq 0)$,

$$\mathbf{x}^T([\mathbf{K}_P][\mathbf{A}])\mathbf{x} = \left(\sqrt{\frac{b_{12} b_{13}}{2 b_{23}}} x_1 + \sqrt{\frac{b_{12} b_{23}}{2 b_{13}}} x_2 + \frac{EF}{2D} \sqrt{\frac{b_{13} b_{23}}{2 b_{12}}} x_3 \right)^2$$
$$+ \left(b_1 - \frac{b_{12} b_{13}}{2 b_{23}} \right) x_1^2 + \left(b_2 - \frac{b_{12} b_{23}}{2 b_{13}} \right) x_2^2 + \left(b_3 - \frac{b_{13} b_{23}}{2 b_{12}} \right) x_3^2$$
$$\tag{7.14}$$

which is strictly positive if

$$b_1 > \frac{b_{12} b_{13}}{2 b_{23}}, \quad b_2 > \frac{b_{12} b_{23}}{2 b_{13}}, \quad b_3 > \frac{b_{13} b_{23}}{2 b_{12}} \tag{7.15a, b, c}$$

Thus, $\mathbf{K}'_P$ is positive definite if $\mathbf{K}_P$ satisfies (7.14) and (7.15). The elements in $\mathbf{K}_D$ can be selected similarly.

7.2.3 DFC with Multi-Sensors

As the DFC method can be applied within any bijective domain, it can be extended to a larger region formed by connecting bijective domains of different sensor sets, where the controlled MFD vector $\mathbf{B_S}$ switches measurements among different sensor sets. Since orientation measurements are not available, alternative MFD-defined Ω domains satisfying the following conditions must be found to determine specified switching points from one domain to the next:

- Ω is completely enclosed by a bijective domain so that bijection holds in Ω.
- Ω boundaries can be defined explicitly in terms of MFDs.

The DFC design concept is best explained with two practical examples. The first is a numerically simulated 1-DOF system illustrating the principle, components and implementation issues involved in a typical DFC system. The second is a practical

7.2 Direct Field-Feedback Control

application built upon an existing 3-DOF ball-joint-like PMSM which provides a basis for experimental investigation of the DFC design principles.

7.3 Numerical 1-DOF Illustrative Example

Figure 7.3a, b illustrates 1-DOF system consisting of a pair of stationary electromagnets (EMs) and a PM free to rotate (angle θ) in the horizontal XY plane. Both the PM and EMs are cylindrical with their magnetization pointing radially. The MFD and torque models are based on closed-form solutions using DMP methods where the detailed computations are given in the appendix. The values of the parameters used in the following simulation are summarized in Table 7.1.

The dynamic equations of the rotor motion and the torque model for this 1-DOF system are given by (7.16):

$$J\ddot{\theta} + b\dot{\theta} = T_Z \quad \text{where} \quad T_Z = [k_{Z1} \quad k_{Z2}]\begin{bmatrix} u_1 \\ u_2 \end{bmatrix} \quad (7.16a,b)$$

In (7.16b), k_{Z1} and k_{Z2} are the torque coefficients TCs (scalars in this case) for the current inputs to EM_1 and EM_2 contributing to the torque T_Z about the Z axis in Fig. 7.3b, which are the components of (7.32) along the Z-axis. Since $N_E = 2$ is greater than the mechanical DOF of one, an optimal current input vector **u** for realizing a desired torque T_{Zd} is given by the inverse torque model for the 1-DOF motion system:

$$\mathbf{u} = \begin{bmatrix} u_1 \\ u_2 \end{bmatrix} = \frac{T_{Zd}}{k_{Z1}^2 + k_{Z1}^2}[k_{Z1} \quad k_{Z2}]^T \quad (7.17)$$

7.3.1 Sensor Design and Bijective Domain Identification

In a DFC system, the rotor MFD is capitalized for sensing feedback proportional to the angular displacement being controlled and for computing the inverse torque model in real-time. Two different sensing design examples (single 1-axis and multi 2-axis sensor) are illustrated:

Sensor Configuration 1: Single 1-axis Sensor
As shown in Fig. 7.3a, a magnetic sensor S_0 measuring B_t (the tangential component of the MFD) is located at the mid-point between the two EMs. Both the **B**-model and torque models can be obtained by means of calibration using the embedded sensor at S_0. Computed from the equations given in the appendix, the relationships among the MFD, TCs and displacement θ are numerically illustrated in Fig. 7.3c–f.

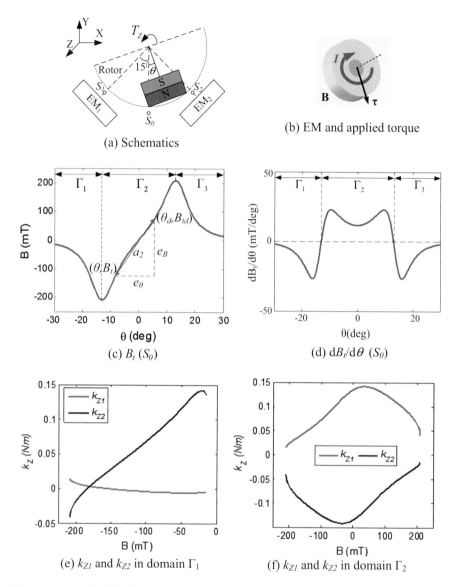

Fig. 7.3 Illustrative 1-DOF motion example

The B_t-θ relationship and its spatial derivative $dB_t/d\theta$ are graphically plotted in Fig. 7.3c, d respectively showing that the range can be divided into three domains (Γ_1, Γ_2, and Γ_3), and that $dB_t/d\theta$ is strictly positive in Γ_2 and negative in Γ_1 and Γ_3. These imply that the curve in each domain is monotonic, and that the B_t-θ relationship in Fig. 7.3c is bijective and can be used to characterize the **B**-model (that determines B_{td} for a given orientation θ_d at S_0).

7.3 Numerical 1-DOF Illustrative Example

Table 7.1 PM and EM parameters for 1-DOF example

PM: [diagram showing PM block with a_o width, magnetization **M** upward, thickness l]   EM: [diagram showing EM block with a_o outer, a_i inner, Core, thickness l]

M: magnetization, $|\mathbf{M}| = M_0$ $\mu_0 = 1.26 \times 10^{-6}\,H/m$

PM	$a_o = 31.75$ mm, $l = 6.35$ mm, $\mu_0 M_0 = 1.465$ T DMP_{PM}: $n = 10$, $k = 4$, $\bar{l}/l = 0.3$ $m_i(\mu A/m)$: 33.5, 24.5, 57.6, 52.0, 276.1
EM	# of turns: 800, $a_o = 31.75$ mm, $a_i = 9.525$ mm, $l = 9.525$ mm DMP_{EM}: $n = 16$, $k = 6$, $\bar{l}/l = 0.442$ $m_i(\mu A/m)$: 1.476, 0.547, 1.618, 1.644, 1.654, 1.325, 0.592

J (kg-m^2)	b (Nm-s/deg)	k_p	k_d
0.02	0.02	0.6	0.04

Similarly, the TCs for computing the inverse torque model (7.17) are functions of $\mathbf{B}(\theta)$. For each θ, the corresponding TCs and B_t can be paired and their profiles in Γ_1 and Γ_2 are graphed in Fig. 7.3e, f respectively; the TCs in Γ_1 and Γ_3 are symmetric. It can be seen that the TCs can be represented as functions of B_t.

Sensor Configuration 2: Three 2-axis Sensors
As bijections are not satisfied at the domain boundary ($dB_t/d\theta = 0$), a multi-sensor configuration is required with overlapping bijective domains to cover the entire workspace without any singularity.

As an illustration, three two-axis magnetic sensors (S_0, S_1, S_2) are located as shown in Fig. 7.4a measuring both the tangential component B_t and the normal component B_n of the rotor MFD, which define the forward **B**-Model and switching criteria respectively. Figures 7.4a, b depict simulated B_t and B_n at S_0; and Fig. 7.4c shows the rotor tangential MFD of three sensors in the bijective domain Γ_2. As the bijective domains (Γ's) overlap eliminates the singularity at the switching point, alternative boundaries for the bijective domains are defined in terms of B_n and denoted as Ω_1, Ω_2 and Ω_3 in Fig. 7.4c. These B_n-defined boundaries are applied as criteria for the entire range of the DFC control law to allow switching **S** (and hence $\mathbf{B}_S$ and $\mathbf{B}_{Sd}$) from one domain to the next.

7.3.2 Field-Based Control Law

With direct MFD feedback, the DFC control law takes the form given in (7.15) where B_t is the tangential MFD:

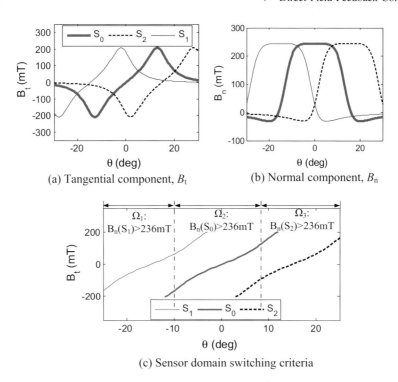

Fig. 7.4 Domain identification of three 2-axis magnetic sensors

$$T_{Zd} = k_p e_B + k_d \dot{e}_B, \quad \text{where } e_B = B_{td} - B_t \tag{7.18a, b}$$

The error e_B can be interpreted as two different points on the B_t-curve in Fig. 7.3c within a bijective domain:

$$a(\theta) = \frac{B_{td} - B_t}{\theta_d - \theta}; \quad \text{hence, } e_B = a e_\theta. \tag{7.19}$$

Apparently, a is a function of θ and has the same sign as $dB_t/d\theta$. Substituting (7.19) into (7.18a) yields the auxiliary control law in a standard form for an orientation-feedback PD control:

$$T_{Zd} = (a k_p) e_\theta + (a k_d) \dot{e}_\theta \tag{7.20}$$

Equation (7.20) implies the 2nd order dynamic system (7.16a) is stable as long as ak_p and ak_d are positive. Although a is not a constant as the states vary, it is bounded by $dB_t/d\theta$ and thus is positive in all three Ω domains as shown in Fig. 7.4c.

7.3.3 Numerical Illustrations of Multiple Bijective Domains

To show the concept feasibility of multiple bijective domains, the smoothness of the domain switching in Sensor Configuration 2 can be numerically examined using the following trajectory on the DFC system:

$$\theta_d = \theta_0 \sin(\pi t/2), \quad \text{where } \theta_0 = 14° \quad (7.21)$$

A portion of each sensor bijective domain as shown in Fig. 7.4c is selected for feedback.

The simulated results are summarized in Fig. 7.5. Figure 7.5a compares the desired and simulated orientation (computed off-line for illustration since the DFC system does not require explicit orientation feedback). Similarly, the reference B_{td} and simulated B_t (corresponding to θ_d and θ) along with the sensor switching sequence are compared in Fig. 7.5b. The direct B_t feedback enables the real-time computation of TCs (k_{Z1} and k_{Z2}) and current inputs (u_1 and u_2) from the inverse model (7.16b); the maximum currents are less than 0.4 A. With the computed TCs and current inputs, the actual torque applied on the rotor is compared against the desired torque determined by the control law (7.20) in Fig. 7.5d. It can be seen that the trajectories of the controlled $\theta(t)$, simulated $B_t(t)$, computed $T(t)$ and $u(t)$ are numerically smooth.

7.4 Experimental Investigation of DFC for 3-DOF PMSM

For investigating the trajectory control performance, the DFC method has been experimentally implemented on a ball-joint-like PMSM as shown in Fig. 7.6. Practical issues encountered in its implementation on a three-DOF orientation controlled system are addressed and discussed. Specifically, this section will begin with a system description for the PMSM, which provides a platform for investigating the bijective relationship (between the MFDs and orientation of the rotor), the computation of the TCVs, and the trajectory performance of three-DOF orientation control in the following subsections.

7.4.1 System Description

Figure 7.6 shows the prototype three-DOF PMSM used in the experimental investigation of the DFC orientation system. As shown in the CAD model in Fig. 7.6a, the PMSM consists of 24 cylindrical EMs housed on the outer surface of the ball-like stator which concentrically supports the socket-like rotor (where 24 cylindrical PMs with alternating polarities are embedded on its inner surface) by

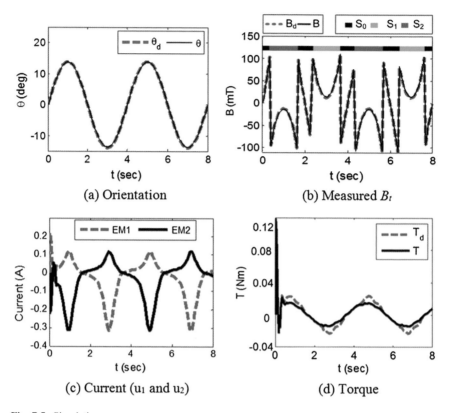

Fig. 7.5 Simulation responses

means of a low-friction spherical bearing. The magnetization axes of both stator EMs and rotor PMs radially pass through the spherical center.

The geometrical centers of the ith PM and EM are expressed in spherical coordinates (r, ϕ, θ) defined in Fig. 7.6b and (7.22):

$$\mathbf{C}_i = R_i [\, C_{\phi i} S_{\theta i} \quad S_{\phi i} S_{\theta i} \quad C_{\theta i} \,]^\mathrm{T} \tag{7.22}$$

where C and S represent cosine and sine of the angles (ϕ or θ) written as a subscript respectively. The 24 EMs are grouped into a total of 12 input pairs (symmetrical about the motor center) in series as shown in Table 7.2, where the rotor orientation is defined in terms of Euler angles $\mathbf{q} = [\,\alpha \quad \beta \quad \gamma\,]^\mathrm{T}$. The locations and indexes of PMs and EMs are given in Table 7.2, where the motion range and inertia properties of the rotor are also listed for completeness.

A weight-compensating regulator (WCR, introduced in Sect. 3.5.3 in Chap. 3) consisting of two circular PM rings (RI and RII) supports the rotor of the PMSM

7.4 Experimental Investigation of DFC for 3-DOF PMSM

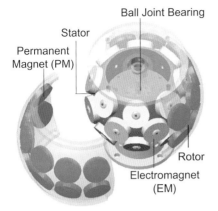

(a) CAD model of a PMSM

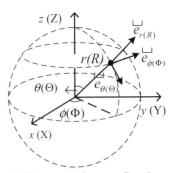

XYZ: stator frame, fixed
xyz: rotor frame, moving

(b) Coordinate systems

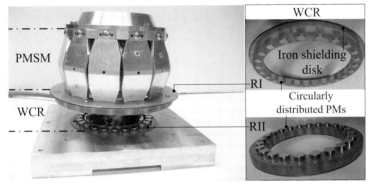

(c) PMSM with Weight-compensating regulator (WCR)

(d) Stator with embedded sensors

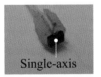

(e) Single-axis

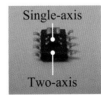

(f) Modified 3-axis

Fig. 7.6 Prototype PMSM

Table 7.2 System parameters

Component	Properties			
ith PM $R_i = 67.9$ mm (Table 7.1)	Index	(θ, ϕ) location in xyz frame		Polarity
	1–12	105°	$(i - 1) \times 30°$	$(-1)^i$
	13–24	75°	$(i - 13) \times 30°$	$(-1)^{i-1}$
jth EM $R_j = 56.8$ mm (Table 7.1)	Index	(Θ, Φ) location in XYZ frame		
	1–8	116°	$(j - 1) \times 45°$	
	9–16	64°	$(j - 9) \times 45°$	
	17–24	0°	$(j - 17) \times 45° + 22.5°$	
Current inputs	$i_1 = i_{13} = u_1$ $i_2 = i_{14} = u_2$ $i_3 = i_{15} = u_3$ $i_4 = i_{16} = u_4$	$i_5 = i_9 = u_5$ $i_6 = i_{10} = u_6$ $i_7 = i_{11} = u_7$ $i_8 = i_{12} = u_8$	$i_{17} = i_{21} = u_9$ $i_{18} = i_{22} = u_{10}$ $i_{19} = i_{23} = u_{11}$ $i_{20} = i_{24} = u_{12}$	
Rotor	Range of motion: $-22.5° \leq \alpha, \beta \leq 22.5°, -\infty \leq \gamma \leq +\infty$			
	Mass (kg): 1.99			
	Moment of inertia (kg-m^2): $I_{xx} = I_{yy} = 6.26 \times 10^{-3}$, $I_{zz} = 8.23 \times 10^{-3}$			

angularly against gravity (Fig. 7.6c). The two PM rings of the WCR are designed to have like polarities facing each other exerting repulsive forces to maintain the rotor at its equilibrium (z- and Z-axis aligned) when **u** = 0.

7.4.2 Sensor Design and Bijective Domains

The multi-sensor system as illustrated in Fig. 7.7 has been designed for the DFC system of the PMSM with the following considerations:

Sensor Layout and Configurations

(a) Since the magnetic structure is symmetric, the sensors for measuring rotor MFD are only placed in half of the stator.
(b) The physical space between the rotor and stator, along with the fact that typical off-the-shelf three-axis magnetic sensor has a relatively large footprint, limiting the sensor design to two configurations as shown in Fig. 7.6d:
   – Single-axis sensor (Fig. 7.6e) embedded in the EM core.
   – Modified three-axis sensors (Fig. 7.6f) constructed by attaching a single-axis sensor on a two-axis sensor and mounted in space in-between EMs.
(c) Regardless the number of axes on a sensor, each measuring axis of the sensor is treated as an independent sensor component and used as an element in the vector $\mathbf{B_S}$ as shown in Fig. 7.2.

7.4 Experimental Investigation of DFC for 3-DOF PMSM

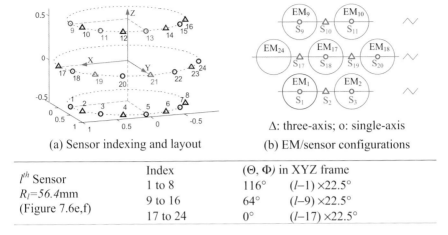

Fig. 7.7 Schematics illustrating sensor layouts

Expressed in spherical coordinates (R, Φ, Θ) defined in Fig. 7.6b, the sensor configurations and locations are shown in Fig. 7.7. The MFD components measured by the sensor and the unit vectors $(\vec{e}_R, \vec{e}_\Theta, \vec{e}_\Phi)$ along the measuring axes of a sensor are defined by (7.23a) and (7.23b) respectively:

$$\mathbf{B} = \begin{bmatrix} B_R & B_\Theta & B_\Phi \end{bmatrix}^T \tag{7.23a}$$

where

$$\vec{e}_R = \begin{bmatrix} S_\Theta C_\Phi \\ S_\Theta S_\Phi \\ C_\Theta \end{bmatrix}; \quad \vec{e}_\Theta = \begin{bmatrix} C_\Theta C_\Phi \\ C_\Theta S_\Phi \\ -S_\Theta \end{bmatrix}; \quad \vec{e}_\Phi = \begin{bmatrix} -S_\Phi \\ C_\Phi \\ 0 \end{bmatrix} \tag{7.23b}$$

For the PMSM prototype in Fig. 7.6, the *Allegro A1302* and *Melexis MLX91204* are used for the single-axis and modified three-axis Hall-effect sensors.

7.4.3 Bijective Domain

To determine the bijective domains relating $\mathbf{B}_S$ and $\mathbf{q}$, the MFDs of the rotor PMs at the sensor locations are computed as functions of rotor orientation using the DMP model (A1). Due to the periodicity of the rotor PM placement, bijective domains of specified sensors are numerically determined using (7.13) for the following region Λ (Fig. 7.8a):

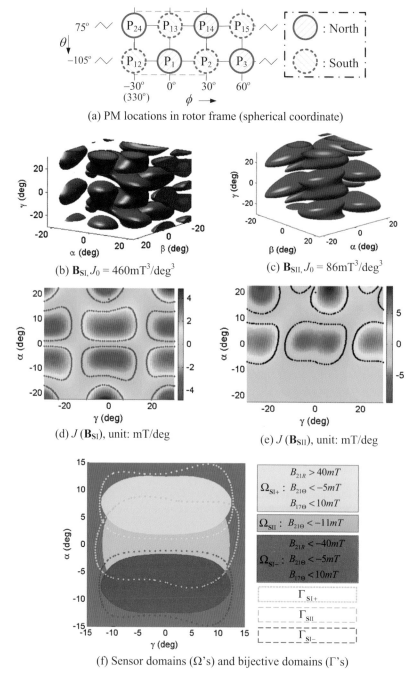

Fig. 7.8 Illustration of sensor bijective domains

7.4 Experimental Investigation of DFC for 3-DOF PMSM

$$\Lambda: -22.5° \leq \alpha, \beta \leq 22.5°, -30° \leq \gamma \leq 30°$$

The results are then extended to the entire working space. The specific sensors in a field feedback vector $\mathbf{B_S}$ (Fig. 7.2) are chosen for the following considerations:

(a) The field-based sensor must be designed such that the feedback $\mathbf{B}_s$ at any orientation outputs three elements, each of which corresponds to a *xyz* Euler angle.
(b) As illustrated in Fig. 7.7a, $SP_{XZ} = \{B_{9R}, B_{17R}, B_{1R}\}$, $SP_{YZ} = \{B_{13R}, B_{21R}, B_{5R}\}$ lie on two orthogonal planes, and are most sensitive to rotations about the *x*, *y* axes respectively. $SP_{mid} = \{B_{11R}, B_{19R}, B_{3R}\}$ is in the middle of SP_{XZ} and SP_{YZ} and are sensitive to rotation about *z*-axis while less sensitive to the other two rotations.
(c) To ensure the entire operating range is bijective, redundant sensors are added on two planes parallel to SP_{XY} such that $SP_{XY-} = \{B_{1R}, \ldots, B_{8R}\}$ and $SP_{XY+} = \{B_{9R}, \ldots, B_{16R}\}$. Sensors on SP_{XY} and $SP_{XY\pm}$ result in 8 planes perpendicular to SP_{XY}; $SP_{\perp i} = \{B_{iR}, B_{(16+i)R}, B_{(8+i)R}\}$, among them $SP_{\perp 1} = SP_{XZ}$ and $SP_{\perp 5} = SP_{YZ}$. Thus, $\mathbf{B_S}$ can be simply formed by selecting the 1st and 2nd sensor readings from $SP_{\perp i}$ and $SP_{\perp (i+5)}$ (which are orthogonal) respectively, and the third from the plane in the middle which is sensitive to rotations about the *z*-axis.
(d) To avoid numerical errors, a relatively large critical value ε in (7.9a) is set so that bijective domains overlap each other; for this reason, the mean absolute value J_0 within the region Λ is chosen as the critical value ε in (7.9a).

<u>Numerical example illustrating connected bijective domains</u>
Two sensor vectors $\mathbf{B_{SI}}$ and $\mathbf{B_{SII}}$ in (7.23a) and (7.23b) with indices defined in Fig. 7.7 are chosen to illustrate the three-DOF orientation about the equilibrium (0, 0, 0):

$$\mathbf{B_{SI}} = (B_{21R}, B_{9R}, B_{19R}); \quad \text{and} \quad \mathbf{B_{SII}} = (B_{13R}, B_{9R}, B_{11R}) \quad \text{(7.24a, b)}$$

Figure 7.8b, c show the bijective domains of the two feedback vectors $\mathbf{B_{SI}}$ and $\mathbf{B_{SII}}$. To facilitate discussions, Fig. 7.8d, e graph the Jacobians for $\mathbf{B_{SI}}$ and $\mathbf{B_{SII}}$ at $\beta = 0$ respectively showing that their bijective domains correspond to different regions in Λ and can be connected to form a larger bijective domain (as shown in Fig. 7.8f) where the DFC can be applied. Meanwhile, alternative MFD-defined domains can be found and Fig. 7.8f depicts the boundaries of the bijective domains (Γ's) and the sensors domains (Ω's). Table 7.3 summarizes the ranges of $\partial B_i/\partial q_j$ within each Ω domain. For simplicity in visual illustration, the MFD and bijective domains were only graphed in the range: $|\alpha|$ and $|\gamma| \leq 15°$ and $\beta = 0$. Note that $\mathbf{B_{SI}}$ has two isolated bijective/sensor domains, which are denoted by "+" and "−" corresponding to $\alpha > 0$ and $\alpha < 0$ respectively. Thus, the DFC method can be applied in these connected domains using the criteria defined by the boundary conditions as shown in Fig. 7.8f to switch the controlled MFD from $\mathbf{B_{SI}}$ to $\mathbf{B_{SII}}$ and vice versa.

Table 7.3 Element value ranges of the Jacobian matrices

Jacobian matrices	Ranges of elemental values (mT/deg), $J(a, b) = \partial B_a/\partial b$		
Ω_{SI+}	$1.1 \leq J(21R, \alpha) \leq 10.9$	$-1.6 \leq J(21R, \beta) \leq 1.6$	$-17.3 \leq J(21R, \gamma) \leq 17.3$
	$-2.8 \leq J(9R, \alpha) \leq 2.8$	$-6.1 \leq J(9R, \beta) \leq -4.3$	$-10.9 \leq J(9R, \gamma) \leq 10.9$
	$-5.8 \leq J(19R, \alpha) \leq 5.8$	$-5.9 \leq J(19R, \beta) \leq 5.9$	$4.0 \leq J(19R, \gamma) \leq 15.9$
Ω_{SI-}	$1.1 \leq J(21R, \alpha) \leq 10.9$	$-1.6 \leq J(21R, \beta) \leq 1.6$	$-17.3 \leq J(21R, \gamma) \leq 17.3$
	$-2.8 \leq J(9R, \alpha) \leq 2.8$	$-6.1 \leq J(9R, \beta) \leq -4.3$	$-10.9 \leq J(9R, \gamma) \leq 10.9$
	$-5.8 \leq J(19R, \alpha) \leq 5.8$	$-5.9 \leq J(19R, \beta) \leq 5.9$	$-15.9 \leq J(19R, \gamma) \leq -4.0$
Ω_{SII}	$-6.1 \leq J(13R, \alpha) \leq -2.1$	$-3.5 \leq J(13R, \beta) \leq 3.5$	$-21.6 \leq J(13R, \gamma) \leq 21.6$
	$-2.7 \leq J(9R, \alpha) \leq 2.7$	$-6.1 \leq J(9R, \beta) \leq -2.1$	$-10.6 \leq J(9R, \gamma) \leq 10.6$
	$-3.8 \leq J(11R, \alpha) \leq 4.9$	$-3.9 \leq J(11R, \beta) \leq 4.8$	$-17.7 \leq J(11R, \gamma) \leq -3.9$

7.4.4 TCV Computation Using Artificial Neural Network (ANN)

The TCV of an EM has a direct relationship with the rotor MFDs enclosing the EM. Although there is no explicit model for computing the TCV from the scattered **B** measurements, with the aid of (7.31) and (7.32) in the appendix a direct mapping between the measured MFD and the TCV can be trained using an ANN. For computing the TCV of the jth EM ($\mathbf{K}_j$), seven measured **B** components from three sensor locations (a single-axis and two three-axis Hall-effect sensors) closest to the jth EM are employed as inputs to the ANN. With the trained ANN, the TCV can be estimated directly using measured MFD and the estimation only requires arithmetic operations in real-time.

Figure 7.9a provides an illustrative example, where an ANN (with 1 hidden layer and 10 nodes) was trained (using 16,200 computed input-output training samples) to map seven measured **B** components at three sensor locations (S_{17}, S_{18} and S_{19}), which are closest to EM_{17} (Fig. 7.7b), onto the three components of $\mathbf{K}_{17}$. The ANN-estimated $\mathbf{K}_{17}$ are compared against results computed analytically using (A3) in Fig. 7.9b as the rotor follows a trajectory given by (7.25):

$$\alpha = 10° \sin t, \quad \beta = 5° \sin t, \quad \gamma = 5°, \quad t \in [0, 2\pi] \tag{7.25}$$

Note that the TCV is orientation dependent; and that the results show excellent agreement.

7.4.5 Experimental Investigation

To examine the DFC concept feasibility, an experimental setup (Fig. 7.10) was designed to command the end effector of the prototype PMSM to continuously track a series of desired orientations. To help visual illustration, a laser pointer (with

7.4 Experimental Investigation of DFC for 3-DOF PMSM

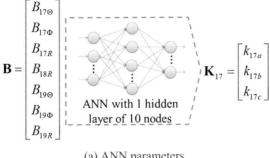

(a) ANN parameters

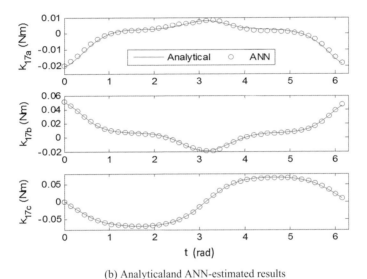

(b) Analyticaland ANN-estimated results

Fig. 7.9 Illustration of TCV estimation with ANN

a gyroscope that records the orientation for experimental verification) is mounted on the rotor for projecting a beam (point **P**) on the planar screen as illustrated in Fig. 7.10a, b, where a camera is placed on the other side of the screen to capture the projection trajectory.

In Fig. 7.10b, the screen coordinate system uv is parallel to the XZ and camera sensor planes; all three planes are perpendicular to the line connecting the rotor center and the origin of the uv coordinate system; D is the distance between the screen and the XZ plane; the laser beam is parallel to the y axis of the rotor frame; and h is the distance from **O** to **C** (interception of the laser beam and z axis). When the rotor orientation is $(0, 0, 0)$, the coordinate of the projection of the laser beam with respect to the uv frame is $(u, v) = (0, h)$.

For any point (X, Y, Z) with respect to the stator frame, the line equation of the laser beam (that passes **C** and is parallel to y-axis) has the form:

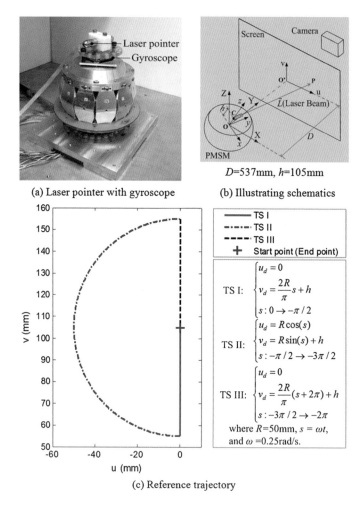

Fig. 7.10 Trajectory control experiment setup

$$\frac{X - hS_\beta}{-S_\gamma C_\beta} = \frac{Y + hC_\beta S_\alpha}{C_\gamma C_\alpha - S_\gamma S_\beta S_\alpha} = \frac{Z - hC_\beta C_\alpha}{C_\gamma S_\alpha + S_\gamma S_\beta S_\alpha} \quad (7.26)$$

The projection which is the interception of the laser beam and the screen can be found by substituting the equation of the screen ($Y = D$) to (7.26), which have the form:

$$\begin{cases} X = -\frac{(D + hC_\beta S_\alpha)S_\gamma C_\beta}{C_\gamma C_\alpha - S_\gamma S_\beta S_\alpha} + hS_\beta \\ Y = D \\ Z = \frac{(D + hC_\beta S_\alpha)(C_\gamma S_\alpha + S_\gamma S_\beta S_\alpha)}{C_\gamma C_\alpha - S_\gamma S_\beta S_\alpha} + hC_\beta C_\alpha \end{cases} \quad (7.27)$$

7.4 Experimental Investigation of DFC for 3-DOF PMSM

In the XYZ coordinate, the uv coordinates have the form: $u = X$, and $v = Z$.

Figure 7.10c shows the desired trajectory (a closed curve consisting of a semi-circular arc and a diameter) on the planar screen; expressed in parametric form, the trajectory is divided into three sections (TS-I, TS-II and TS-III). As the rotor has 3-DOF orientation while the trajectory is defined on a two-dimensional plane, the unique solutions with a constraint $\beta = 0$ can be obtained in this experimental task.

As illustrated above, the major components of the DFC system are derived and analyzed using the DMP and dipole force models given in the appendix with the exception of the forward **B**-Model (relating $\mathbf{q}_d$ to $\mathbf{B}_{Sd}$) obtained through calibration to compensate for manufacturing tolerances.

Calibration

As assumed in the derivation, the DFC bases solely on rotor MFD in the feedback loop. Physical MFDs are contributed by both the rotor PMs and energized EMs; the latter (proportional to its current) must be subtracted off from the real-time measurements ($B_{p,meas}$) of the pth sensor component B_p:

$$B_p(\alpha, \beta, \gamma) = B_{p,meas}(\alpha, \beta, \gamma) - \sum_{j=1}^{N_E} c_{j,p} u_j \quad (7.28)$$

The calibration process includes two independent steps:

Step 1: Acquire $B_{p,meas}$ as a function of Euler angles when no current is supplied.
Step 2: As the MFD of an EM is independent of rotor orientation, the constant coefficients relative to the current input can be calibrated at a fixed rotor orientation or with the absence of the rotor PMs.

Figure 7.11 shows the experimental setup for acquiring $B_{p,meas}$, where the shaft of the PMSM is positioned by two rotary guides (each driven by a stepper motor with a timing belt) about the X and Y axes. The spin of the shaft about the z-axis is independently driven by the third step-motor. An overall 10:1 step ratio was used all three axes to achieve rotational resolutions of 0.54°/step about the X and Y axes, and 0.18°/step about the z-axis. During calibration, the MFDs for a specified incremental orientation are measured and stored in a 3D table for interpolation in real-time.

In (7.28), the proportionality constant $c_{j,p}$ (that represents the ratio of measured B_p component at S_p over the jth current input) can be calibrated before the rotor PMs is installed as shown in Fig. 7.7a, where the embedded sensors measure the MDFs of the EMs. As an illustration, consider Fig. 7.7b where the MDF of the EM_{17} is measured by the embedded sensors surrounding it. Figure 7.12 shows the typical $c_{j,p}$ calibration results when the current input u_9 (varying from −1A to 1A) flows through EM_{17} and EM_{21} (in series).

The effects of MFD generated by an EM on a sensor component can be observed in Fig. 7.12:

Fig. 7.11 Setup for 3D calibration

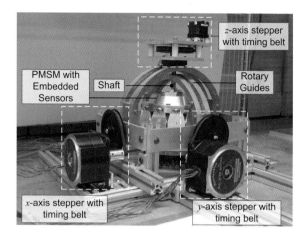

Fig. 7.12 Measured B-components (u_9 from -1A to 1A)

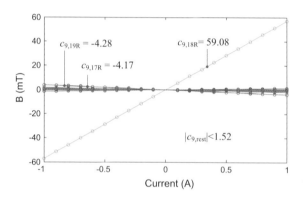

- The MFD readings are linearly proportional to the current input as expected. The calibrated $c_{j,p}$ constants are thus simply the slopes of the lines.
- The MFD of an EM has a dominant effect on the B_R sensor along its core; $c_{9,17}$ and $c_{9,19}$ are about 7% of $c_{9,18R}$ and the remainders are less than 3% of $c_{9,18R}$.

In real-time DFC of the PMSM, $c_{j,p}$ with an absolute value less than 0.5 are neglected to reduce real-time computations.

DFC Implementation

The control objective is to track a desired trajectory with the laser beam. Given the desired orientations, the corresponding desired MFDs are interpolated from the calibrated **B**-model as discussed above. Throughout the entire trajectory, the controlled MFD vector switches between two MFD vectors given in (7.24a, b) among the three sensor domains based on the switching criteria (or boundary conditions) in Fig. 7.8f.

7.4 Experimental Investigation of DFC for 3-DOF PMSM

Table 7.4 Control gain matrices

	K_P	K_I	K_D
Ω_{SI+}	diag(2, −0.8, 1.6)	diag(0.03, −0.02, 0.04)	diag(60, −27, 84)
Ω_{SI-}	diag(2, −0.8, −1.6)	diag(0.03, −0.02, −0.04)	diag(60, −27, −84)
Ω_{SII}	diag(−2, −1, −2)	diag(−0.006, −0.03, −0.05)	diag(−66, −22.5, −54)

As there is a restoring torque from the WCR when the rotor is not at its equilibrium, an integral term is incorporated in the DFC control law (7.10). Note that the integral torque must be accumulated when the control law switches from different sensor sets. The PID gain matrices (Table 7.4) were determined using (7.14) and (7.15) where the maximum and minimum values of the elements J_{ij} within each domain, which represent the bound of a_{ij} in (7.11) can be found in Table 7.3. The DFC system is implemented on a *NI cRIO-9025* real-time controller featured with 800 MHz processor and 512 MB ram.

Results and Discussion

The DFC trajectory tracking results of the PMSM are summarized in Figs. 7.13, 7.14. Figure 7.13 (where the color bars denote the switching sequence) compares the reference and actual controlled MFD vectors, and the projected beam trajectories. The corresponding current inputs are given in Fig. 7.14. As a basis for comparison, the desired Euler angles are compared against those measured by the gyroscope in Fig. 7.15. The relative errors (with respect to the difference between maximum and minimum values, or peak to peak values) summarized from the results graphed in Figs. 7.13 and 7.14 are listed in Table 7.5, where the relative projected error E_p is defined in (7.29):

$$E_p = \begin{cases} (u_d - u)/R, & \text{TS-I} \\ \left(R - \sqrt{u^2 + (v-h)^2}\right)/R, & \text{TS-II} \\ (u_d - u)/R, & \text{TS-III} \end{cases} \quad (7.29)$$

In (7.29), u_d and the trajectory sections are defined in Fig. 7.10c.
Some observations can be made from the above results:

- As shown in Fig. 7.13a, all three components of the controlled variable follow those of the reference MFD very closely with a mean relative error of less than 0.5% and time delay of 12 ms (as shown in Fig. 7.13b). These results are consistent with the comparison shown in Fig. 7.13c where the coordinates of the projected laser beam captured by the camera closely matches the desired reference path on the screen.
- As shown in Fig. 7.14 where the switching sequence (color bar) is superimposed, no oscillation was observed during switching. The smoothness of the switching among the sensor domains is also confirmed by the continuously changing inputs in Fig. 7.15 with a maximum current of less than 0.5 A.

148   7 Direct Field-Feedback Control

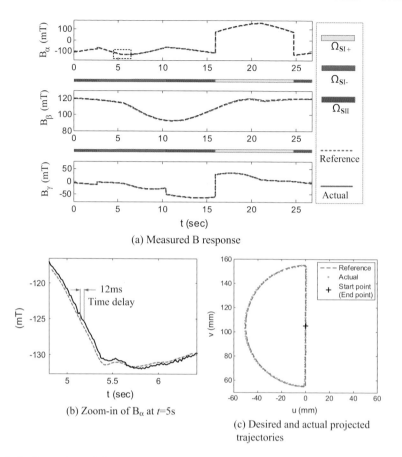

Fig. 7.13 Comparison of desired and actual responses

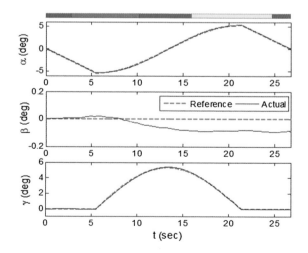

Fig. 7.14 Orientation response

7.4 Experimental Investigation of DFC for 3-DOF PMSM

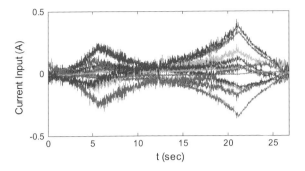

Fig. 7.15 Current inputs

Table 7.5 Percentage error

	Magnetic sensors			Gyroscope			Image
	B_α	B_β	B_γ	α	β	γ	E_p
|A|	285.3 mT	153.2 mT	99.2 mT	10.6°	0°	10.6°	50 mm
Max (%)	1.21	0.95	0.75	1.39	N.A	0.81	1.76
Mean (%)	0.41	0.47	0.31	0.39	N.A	0.39	0.64

\|A\| = Peak to peak difference

- The DFC method eliminates a need for an external sensor such as a gyroscope which requires a sampling time of 12 ms for orientation estimation while the DFC system takes only a sampling time of 4 ms for the closed loop.

In summary, the DFC method for multi-DOF orientation control of a PMSM has been presented, which feedbacks the measured MFD (uniquely corresponding to the rotor orientation) eliminating the need for explicit orientation sensing. Direct field feedback permits parallel computation of the control law and the TCVs, which reduces accumulative errors and time-delay in serial computations commonly found in existing control methods relying on orientation-dependent models.

The method for designing a DFC system has been illustrated with two examples; a one-DOF rotational system and a three-DOF PMSM. Using analytical magnetic field and torque models, the relationship between rotor orientation and MFD are non-unique, which can be overcome by the use of connected bijective domains of a multi-sensor system covering the entire working range. The DFC system has been experimentally implemented on the PMSM test-bed; the results show that the DFC system can offer smooth and precise controlled responses.

Although developed in the context of rotational motion of a PMSM, this approach can be extended to other PM-based motions systems like linear motors. It is expected that the methods developed here for analyzing bijective domains as well as directly estimating the TCV can also benefit the studies of orientation sensing and force/torque control for many different PMSM configurations.

Appendix

Mathematically, an axially magnetized cylindrical PM or EM can be modeled as a distributed set of dipoles as suggested in (7.30), which consists of k circular uniformly spaced loops of n equally spaced dipoles parallel to the magnetization axis:

$$\bar{a}_j = \frac{a_o j}{k+1} \text{ at } z = \pm \frac{\bar{\ell}}{2} \; (0 \leq j \leq k) \tag{7.30}$$

The MFD generated by a PM or EM at point **P** in space is given by (7.31):

$$\mathbf{B} = \frac{I\mu_0}{4\pi} \sum_{i=0}^{k} m_i \sum_{j=1}^{n} \left(\frac{\mathbf{R}_{ij+}}{|\mathbf{R}_{ij+}|^3} - \frac{\mathbf{R}_{ij-}}{|\mathbf{R}_{ij-}|^3} \right) \tag{7.31}$$

where $\mathbf{R}_{ij+}$ and $\mathbf{R}_{ij-}$ are the vectors from the source and sink of the jth dipole on the ith loop to **P** respectively; m_i is the pole strength of the poles on ith loop; for an EM, I is the supplied current; for a PM, $I = 1$ and m_i is a function of the residual magnetization M_0. The detailed process for deriving $\vec{l}, n, k, m_i$ of a PM and an EM can be found in Sects. 3.1 and 3.2 in Chap. 3.

Given that the EM and the entire rotor PMs are characterized by n_s and n_r dipoles respectively, the TCV defined in (7.4a) can be derived using the dipole force method:

$$\mathbf{K}_j = \frac{\mu_0}{4\pi} \sum_{i=1}^{n_r} m_{ri} \sum_{p=1}^{n_s} m_{sp} \left[(\mathbf{R}_{ri+sp+} - \mathbf{R}_{ri+sp-}) \times \mathbf{R}_{ri+} - (\mathbf{R}_{ri-sp+} + \mathbf{R}_{ri-sp-}) \times \mathbf{R}_{ri-} \right] \tag{7.32}$$

where

$$\mathbf{R}_{ri\pm sp\pm} = (\mathbf{R}_{ri\pm} - \mathbf{R}_{sp\pm}) / |\mathbf{R}_{ri\pm} - \mathbf{R}_{sp\pm}|^3. \tag{7.33}$$

In (7.32), $\mathbf{R}_{ri\pm}(\mathbf{R}_{sp\pm})$ and $m_{ri}(m_{sp})$ are the location and pole strength of the ith (pth) dipole pair of the rotor PM (stator EM), where "+" and "−" stand for the source and the sink of the dipole respectively.

References

1. K. Bai, K.M. Lee, Direct field-feedback control of a ball-joint-like permanent-magnet spherical motor. IEEE/ASME Trans. Mechatron. **19**(3), 975–986 (2014)
2. S.G. Krantz, H.R. Parks, *The Implicit Function Theorem: History, Theory, and Applications*, 1st edn.(Birkhäuser Boston, 2002)
3. C.H. Edwards, D.E. Penney, *Multivariable Calculus*, 6th edn. (Prentice Hall, 2002)

Chapter 8
A Two-Mode PMSM for Haptic Applications

Haptic devices, which have the capabilities to provide realistic force/tactile feedback to human operators in a virtual environment, play an increasingly important role in design and training stages in many fields. With the rapid development of computer technology and mechatronics, novel applications of haptic devices can be found in both traditional and emerging industries including robotic automation, medical operations. A PMSM can be used not only as a device to provide smooth and precise multi-DOF motion, but also a manipulator capable of force/torque manipulation due to the direct-drive property. As an immediate application, a ball-joint-like PMSM is presented in this chapter as an alternative design for haptic devices [1]. With a two-mode configuration, this device can be operated as a joystick manipulating a target in six degrees-of-freedom (DOF), and provides realistic force/torque feedback in real-time.

8.1 Description of the PMSM Haptic Device

In Fig. 8.1, the radial magnetization axes (of both the PMs and the air-cored EMs) point towards the joint center; and the entire structure (except for the PMs) is non-magnetic. In spherical coordinates (ϑ, φ, r), the magnetization axes can be characterized by a vector pointing from the origin to the centroid of each PM or EM as shown in Fig. 8.2a. The complete description of PM and EM positions is given in the Appendix. Figure 8.2b illustrates the coordinate systems of a PMSM, where XYZ is the stator frame (stationary); xyz is the rotor frame; and **q** is a vector of XYZ Euler angles describing the rotor orientation:

$$q = [\psi \quad \lambda \quad \phi]^T \tag{8.1}$$

In design and control of a PMSM, both the forward and inverse torque models are needed. The former (used in design analysis) computes the three torque

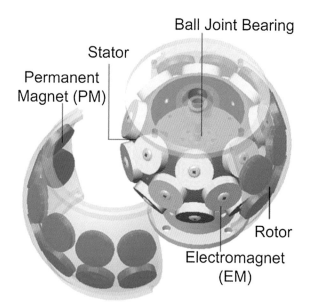

Fig. 8.1 Concept of a PMSM haptic device

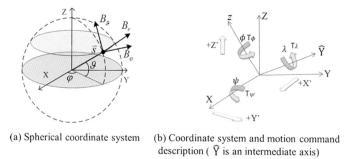

(a) Spherical coordinate system

(b) Coordinate system and motion command description ($\widehat{Y}$ is an intermediate axis)

Fig. 8.2 Illustration of position and torque commands

components for a specified set of electrical currents. Unlike the forward model where the solutions are unique, there are infinite solutions to the inverse torque model (that is required in real-time control) as the number of current inputs is more than three (the number of desired DOF or torque constraint equations). With linear magnetic properties, the electromagnetic torque of the PMSM has the form:

$$\mathbf{T} = [T_\psi \quad T_\lambda \quad T_\phi]^\mathrm{T} = [\mathbf{K}]\mathbf{u} \qquad (8.2)$$

8.1 Description of the PMSM Haptic Device

where

$$[\mathbf{K}] = [\mathbf{K}_1 \quad \ldots \quad \mathbf{K}_j \quad \ldots \quad \mathbf{K}_{Ns}]; \quad \mathbf{u} = [i_1 \quad \ldots \quad i_j \quad \ldots \quad i_{Ns}]^T; \quad (8.3a,b)$$

i_j is the current input to EM_j. In (8.3a), $\mathbf{K}_j$ is the torque coefficient vector (TCV) of EM_j where the TCV is the resultant torque acting on the rotor due to a 1-Ampere current flowing through a single EM. An optimal current vector minimizing the total input energy can be found using Lagrange multipliers:

$$\mathbf{u} = [\mathbf{K}]^T ([\mathbf{K}][\mathbf{K}]^T)^{-1} \mathbf{T} \qquad (8.4)$$

For computing the orientation and TCV, a set of magnetic sensors (given in the Appendix) measuring the magnetic flux density **B** is installed on the stator with their sensor axes pointing in the (ϑ, φ, r) directions for measuring the three components $(B_\varphi, B_\vartheta, B_r)$.

8.1.1 Two-Mode Configuration Design for 6-DOF Manipulation

Physically, the PMSM has three-DOF of rotational motion but can be configured to operate in two modes to achieve two independent sets of (rotational and translational) motion in the target space (as shown in Fig. 8.2b):

- *Rotational mode*: The PMSM can be directly used as an integrated rotational motion-sensor and torque-actuator. The three-DOF rotations are defined as:

$$[\lambda' \quad \psi' \quad \phi']^T = [\varsigma_1 \lambda \quad \varsigma_2 \psi \quad \varsigma_3 \phi]^T \qquad (8.5a)$$

where the constants, ς_1, ς_2 and ς_3, can be tuned to meet specific needs; and the prime denotes the coordinates in virtual environment (similarly hereinafter). Similarly, the PMSM can <u>physically</u> simulate the torque feedback from the <u>virtual</u> target by directly applying Lorenz torques on its rotor in real time enabling the user to have the haptic feel. The three torque components have the form (with constant η_1, η_2 and η_3):

$$[T_\psi \quad T_\lambda \quad T_\phi]^T = [\eta_1 T_{\psi'} \quad \eta_2 T_{\lambda'} \quad \eta_3 T_{\phi'}]^T \qquad (8.5b)$$

- *Translational mode*: The PMSM can also be configured in translational domain such that the user's rotational motion on the rotor is interpreted into translational displacements. By the same token, the force feedback from the virtual target is actuated as torques on the rotor enabling the user to have an equivalent haptic experience:

$$[X'\ Y'\ Z']^{\text{T}} = [\hat{\varsigma}_1 \lambda\ \ \hat{\varsigma}_2 \psi\ \ \hat{\varsigma}_3 \phi]^{\text{T}} \tag{8.6a}$$

$$[T_\psi\ T_\lambda\ T_\phi]^{\text{T}} = [\hat{\eta}_1 F_{Y'}\ \ \hat{\eta}_2 F_{X'}\ \ \hat{\eta}_3 F_{Z'}]^{\text{T}} \tag{8.6b}$$

where $\hat{\varsigma}_1, \hat{\varsigma}_2, \hat{\varsigma}_3$ and $\hat{\eta}_1, \hat{\eta}_2, \hat{\eta}_3$ are constants.

By switching between these two modes, the PMSM is capable of two independent sets of three-DOF motions in the target space providing sensible force/torque feedback to the user in real time.

8.1.2 Numerical Model for Magnetic Field/Torque Computation

As a physical EM or PM (that is axially magnetized and has a cylindrical shape of radius a and length l) can be mathematically characterized with DMP models, the magnetic flux density as well as the magnetic force/torque can be then computed in close-forms. The dipole (with strength m) is defined here as a pair of source and sink separated by a distance $\bar{l}$. This model for an axially magnetized cylindrical PM or EM consists of k circular uniformly spaced loops of n equally spaced dipoles parallel to the magnetization vector:

$$\bar{a}_j = a_o j/(k+1) \quad \text{at} \quad z = \pm \bar{l}/2 (0 \leq j \leq k) \tag{8.7}$$

The flux density generated by a PM or EM at a point in space can be computed using (8.8):

$$\mathbf{B} = \frac{\mu_0}{4\pi} \sum_{i=0}^{k} m_i \sum_{j=1}^{n} \left(\frac{\mathbf{R}_{ij+}}{|\mathbf{R}_{ij+}|^3} - \frac{\mathbf{R}_{ij-}}{|\mathbf{R}_{ij-}|^3} \right) \tag{8.8}$$

where $\mathbf{R}_{ij+}$ and $\mathbf{R}_{ij-}$ are the vectors from the source and sink of the jth dipole on the ith loop to the point being considered, respectively; and m_i is the pole strength of the poles on ith loop.

Given that each of the EM (or PM) is characterized by n_s (or n_r) dipoles, the component $\mathbf{K}_j$ of the EM_j can be derived using the dipole force method:

$$\mathbf{K}_j = \frac{\mu_0}{4\pi} \sum_{i=1}^{N_r \times n_r} m_{r_i} \sum_{p=1}^{n_s} m_{s_p} [(\mathbf{R}_{r_i+s_{p+}} - \mathbf{R}_{r_i+s_{p-}}) \times \mathbf{R}_{r_i+} - (\mathbf{R}_{r_i-s_{p+}} + \mathbf{R}_{r_i-s_{p-}}) \times \mathbf{R}_{r_i-}] \tag{8.9}$$

In (8.9), $\mathbf{R}_{r_{i\pm}s_{p\pm}} = (\mathbf{R}_{r_{i\pm}} - \mathbf{R}_{s_{p\pm}})/|\mathbf{R}_{r_{i\pm}} - \mathbf{R}_{s_{p\pm}}|^3$ where $\mathbf{R}_{r_{i\pm}}(\mathbf{R}_{s_{p\pm}})$ is the ith (pth) pole location of the rotor (EM_j); and the signs, (+) and (−), stand for the source and

8.1 Description of the PMSM Haptic Device

the sink of the dipole respectively; and m_{ri} (m_{sp}) are the pole strength of the ith (pth) dipole pair in the rotor (EM$_j$).

8.1.3 Field-Based TCV Estimation

As shown in (8.8), the magnetic fields of the PMs and hence the TCV of the EMs in (8.9) depend on the rotor orientation. Computing the TCV using orientation-based models would result in significantly long sampling time as the orientation and TCV must be sequentially computed. It is desired that the TCV can be derived directly from magnetic field measurements; the relationship between them can be characterized with a direct mapping as illustrated in Fig. 8.3 where an artificial neural network (ANN) is used to model the complex relationship between the magnetic field and the TCV through the Levenberg-Marquardt supervised back-propagation training algorithm.

As an illustration, consider the TCV ($\mathbf{K}_{17}$) of EM$_{17}$ (located on the X-axis); note that the indexing of the sensors and EMs can be found in the Appendix. The magnetic flux densities as well as the TCV are computed with (8.8) and (8.9) in the working space of the PMSM described as

$$-22.5° \leq (\psi, \lambda) \leq 22.5°, \quad 0° \leq \phi < 360°$$

The parameters used for the computations are described in Table 8.1. An ANN (with 1 hidden layer and 10 nodes) was trained with the computed data (16,200 samples). The inputs, outputs as well as the ANN parameters are shown in Fig. 8.3. As an illustrative comparison, the components of $\mathbf{K}_{17}$ are estimated with the ANN and the $\mathbf{B}$ computed using (8.8) while the rotor follows a trajectory given by

$$\psi = 10° \sin t, \quad \lambda = 5° \sin t, \quad \phi = 5°, \quad t \in [0, 2\pi]$$

The computed results agree excellently well with the analytical solutions computed using (8.9) along the trajectory as compared in Fig. 8.4.

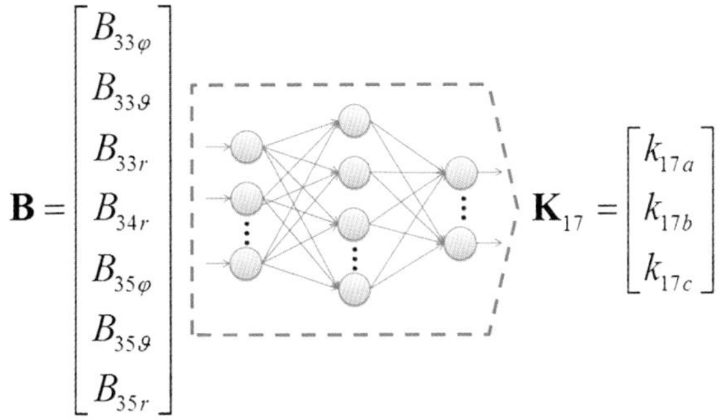

Fig. 8.3 ANN parameters

$$\mathbf{B} = \begin{bmatrix} B_{33\varphi} \\ B_{33\vartheta} \\ B_{33r} \\ B_{34r} \\ B_{35\varphi} \\ B_{35\vartheta} \\ B_{35r} \end{bmatrix} \quad \mathbf{K}_{17} = \begin{bmatrix} k_{17a} \\ k_{17b} \\ k_{17c} \end{bmatrix}$$

Table 8.1 DMP Parameters

PM	EM
$a = 15.875$ mm, $l = 0.2$, $\mu_o M_o = 1.31$ T	$a = 15.88$ mm, $a_r = 0.3$, $l = 0.3$, # of turns = 1000
DMP$_{PM}$: $n = 10$, $k = 4$ $\bar{l}/l = 0.3$	DMP$_{EM}$: n = 16, k = 6, $\bar{l}/l = 0.442$
m_j(μA/m): 33.5, 24.5, 57.6, 52.0, 276.1	m_j(μA/m): 1.48, 0.55, 1.62, 1.64, 1.65, 1.33, 0.60

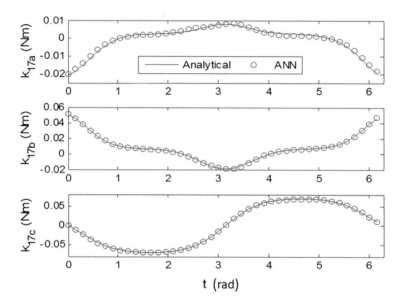

Fig. 8.4 Analytical and ANN-estimated results

8.2 Snap-Fit Simulation

Snap-fits offer resistance to engagement during assembly and disassembly in various types of products, equipment, and power systems. A means to experience the force/torque feedback is desirable for manual assembly in many applications particularly for snap-fit engagement of critical components. Of particular interest here is to explore the application of a PMSM as a haptic device for experiencing the snap-fit assembly/disassembly processes in a virtual environment, and enabling a designer to qualitatively evaluate the level of deformation through realistic force/torque feedback.

Figure 8.5 shows a typical haptic evaluation procedure during the design optimization of a product. Often, several optional configurations for a product are designed considering tradeoffs among many parametric effects (including both geometric construction and material properties). As illustrated in the dashed box labeled as "Optimization," these design configurations are presented in the

8.2 Snap-Fit Simulation

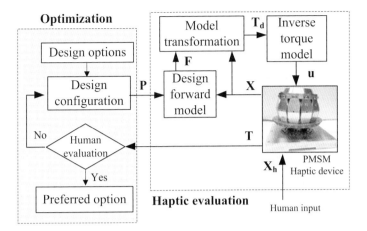

Fig. 8.5 Haptic evaluation procedure and prototype PMSM

parameter vector **P** and output to the "design forward model" along with the vector **X** containing the sensed positional and/or orientation motion corresponding to the human input X_h manipulated by a human designer on the PMSM haptic device. Through the "design forward model", the multi-DOF (manual assembly or disassembly) force/torque vector **F** is then solved in terms of the sensed motion **X** (that is translated into motion command driving the assembly/disassembly component in virtual space). Since PMSM is a 3-DOF angular device, the "model transformation" converts the required force/torque **F** and sensed motion **X** into the desired torque vector T_d for computing an optimal set of current inputs using the inverse torque model (8.4) to drive the spherical rotor. From the magnetic field measurements both the rotor orientation and the TCVs can be determined directly using ANNs, upon which an optimal set of current inputs **u** can be computed. The current inputs (flowing into the EMs in the presence of the rotor PM field) result in feedback torque **T** acting on the rotor while the designer maneuvers the PMSM haptic device. Thus, the designer can then select a preferred configuration based on the "feel" by evaluating the product performance in a virtual environment.

Since the rotor of the PMSM is essentially an inverted pendulum and becomes inherently unstable when no current inputs are supplied, a weight-compensating regulator (WCR) is incorporated in the prototype PMSM. The WCR consisting of two circular PM rings uses a couple of distributed repulsive PM forces to support the rotor angularly against gravity and tends to maintain the rotor at its equilibrium and effectively reduces the input electrical energy required (and thus heat generation). The restoring torque of the WCR (as a function of the inclination angle) can be compensated when generating the torque feedback. Due to the rotor symmetrical PM configuration, the EMs are grouped into 12 electrical inputs (as described in Table 8.2) with each group of two EMs (symmetrical about the motor center) connected in series.

Table 8.2 Current input configuration of the EMs

$i_1 = i_{13} = u_1$	$i_5 = i_9 = u_5$	$i_{17} = i_{21} = u_9$
$i_2 = i_{14} = u_2$	$i_6 = i_{10} = u_6$	$i_{18} = i_{22} = u_{10}$
$i_3 = i_{15} = u_3$	$i_7 = i_{11} = u_7$	$i_{19} = i_{23} = u_{11}$
$i_4 = i_{16} = u_4$	$i_8 = i_{12} = u_8$	$i_{20} = i_{24} = u_{12}$

8.2.1 Snap-Fit Performance Analyses

Figure 8.6 illustrates the snap-fit assembly of a typical cantilever hook [2] (base thickness h_o, width w and length l_t) with a wedge-shaped end characterized by the height h_b and angles (α, β). The shaded cantilever indicates its initial state; and δ is the beam deflection as the matching part contacts the wedge at x. As the matching part advances (or retracts) for assembly (or disassembly), the contact point slides along the front (or rear) surface of the wedge as well as deflects the beam. To offer intuitive insights and facilitate design, the forces and geometrical dimensions are normalized to (Ewh_o) and h_o as follows:

$$\begin{bmatrix} F_x \\ F_y \end{bmatrix} = \frac{1}{Ewh_o} \begin{bmatrix} f_x \\ f_y \end{bmatrix}; \quad X = \frac{x}{h_o}; \quad L_b = \frac{\ell_b}{h_o}; \quad H_b = \frac{h_b}{h_o}; \quad L_t = \frac{\ell_t}{h_o}$$

The normalized beam deflection $\Delta = \delta/h_o$ can be expressed as a mechanical impedance Ω, where the subscripts "+" and "−" indicate the deflections are in the +y and −y directions respectively:

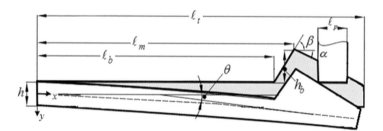

(a) Coordinate systems and characteristic dimensions

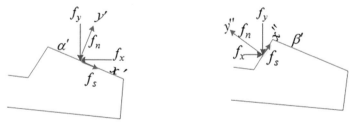

(b) Forces components in insertion   (c) Forces components in retention

Fig. 8.6 Cantilever hook and matching part

8.2 Snap-Fit Simulation

$$\Omega = \Delta/F_y = \Omega_+ + s(f_x/f_y)\Omega_-$$
$$\text{where } s = \begin{cases} +1, & \text{for } x = [\ell_b, \ell_m) \\ -1, & \text{for } x = (\ell_m, \ell_t] \end{cases} \text{ and at } x = \ell_m \quad s = \begin{cases} -1 & \text{assembly} \\ +1 & \text{disassembly} \end{cases}$$
(8.10)

The assembly force f_x and deflecting force f_y are related by (8.11), where μ is the friction coefficient between the two sliding surfaces:

$$\frac{f_x}{f_y} = \begin{cases} \tan(\gamma' + \tan^{-1}\mu) & \text{Insertion} \\ \mu & \text{Dwelling} \\ \tan(\gamma' - \tan^{-1}\mu) & \text{Retention} \end{cases}$$
(8.11)

$$\text{where } \gamma' = \begin{cases} \alpha' = \alpha + \tan^{-1}(\delta/x), & \text{for } x = [\ell_m, \ell_t]; \\ \beta' = \beta - \tan^{-1}(\delta/x), & \text{for } x = [\ell_b, \ell_m]; \end{cases}$$

The normalized deflection ($\Delta_+ = \delta_+/h$) is a function of F_y and can be expressed as

$$\Omega_+ = \frac{\Delta_+}{F_y} = \sum_{i=1}^{3} \Omega_{i+} = 4KL_b^3 + \int_{\ell_b}^{x} \frac{12x^2 dx}{h^3(x)} + \frac{pE}{G}\int_{\ell_b}^{x} \frac{dx}{h(x)}$$
(8.12)

$$\text{where } K = 1 + \frac{0.3}{L_b^2}\left(\frac{E}{G}\right)$$

$$\text{where } h(x) = \begin{cases} h_o, & \text{for } x = [0, \ell_b] \\ h_o + (x - \ell_b)\tan\beta, & \text{for } x = [\ell_b, \ell_m]; \beta = [0, \pi/2) \\ h_o + (\ell_t - x)\tan\alpha, & \text{for } x = [\ell_m, \ell_t]; \alpha = [0, \pi/2) \end{cases}$$

and $h(x) = h_o + h_b$ when $\alpha = \pi/2$ or $\beta = \pi/2$.

Similarly, the normalized deflection ($\Delta_- = \delta_-/h_o$) is a function of F_x:

$$\Omega_- = \frac{\Delta_-}{sF_x} = \Omega_{1-} + \Omega_{2-} = 6X\left(\frac{h(x)}{h_o} - \frac{1}{2}\right)L_b + 6X\left(\frac{h(x)}{h_o} - \frac{1}{2}\right)\int_{\ell_b}^{x} \frac{h_o^2 dx}{h^3(x)}$$
(8.13)

8.2.2 Snap-Fit Haptic Application

Figure 8.7a shows the CAD model of an "outlet" (commonly in household products) demonstrating the contact feeling of a typical cantilever snap-fit during design. The outlet snap-fit consists of a cold-drawn brass plug and a fixed socket made of

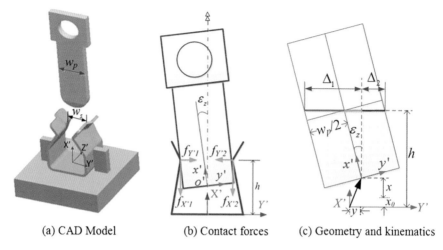

(a) CAD Model   (b) Contact forces   (c) Geometry and kinematics

Fig. 8.7 Schematics illustrating the disassembly ($x_0 = (w_p/2)\sin(\varepsilon_z)$)

rolled phosphor bronze. The extraction of the plug from the socket undergoes the dwelling process of the snap fit as illustrated in Fig. 8.7b.

The disassembly of the outlet snap-fit can be considered as a nonlinear beam deflection problem with the movement of the plug as an input. Since the plug-width is larger than the distance of the initial socket opening, this difference leads to the beam deflection δ. As shown in Fig. 8.7b, the detaching motion of the plug can be characterized as a translation in X direction and rotation about Z-axis (ε_z). The forces applied on the plug can be computed as illustrated in Fig. 8.7b where the net forces ($F_{X'}$, $F_{Y'}$) due to the contact forces are given by (8.14):

$$\begin{bmatrix} F_{X'} \\ F_{Y'} \end{bmatrix} = \begin{bmatrix} f_{X'1} + f_{X'2} \\ f_{Y'1} - f_{Y'2} \end{bmatrix} \tag{8.14}$$

Since the extraction undergoes a dwelling process; thus (8.10) and (8.11) apply with $s = +1$ and $fx/fy = \mu$:

$$\begin{bmatrix} f_{X'1} \\ f_{X'2} \end{bmatrix} = \frac{\mu E w}{\Omega_+ + \mu \Omega_-} \begin{bmatrix} \Delta_1 - w_s/2 - y \\ \Delta_2 - w_s/2 + y \end{bmatrix} \quad \text{and} \quad \begin{bmatrix} f_{Y'1} \\ f_{Y'2} \end{bmatrix} = \frac{1}{\mu} \begin{bmatrix} f_{X'1} \\ f_{X'2} \end{bmatrix} \tag{8.15a, b}$$

In this illustrative example, the plug is initially inserted such that $\varepsilon_z = 0$ (and $x'y'$ coincides with $X'Y'$). The kinematic parameters can be derived with the aid of Fig. 8.7c:

$$\Delta_{1,2} = \frac{w_p}{2 \cos \varepsilon_z} \pm (h - x - x_o) \tan \varepsilon_z \tag{8.16}$$

8.2 Snap-Fit Simulation

Hence, the net forces can be written in closed form given by (8.17):

$$\begin{bmatrix} F_{X'} \\ F_{Y'} \\ F_{Z'} \end{bmatrix} = \frac{Ew}{\Omega_+ + \mu\Omega_-} \begin{bmatrix} \mu w_p \sec\varepsilon_z - \mu w_s \\ 2\tan\varepsilon_z(h - x_0 - x) - 2y \\ 0 \end{bmatrix} \quad (8.17)$$

where

$$\Omega_+ = \sum_{i=1}^{3} \Omega_{i+} \quad \text{and} \quad \Omega_- = \sum_{j=1}^{2} \Omega_{j-}$$

$$\rho(h_o) = \frac{1}{1 + H_b}; \quad H_x = 1 + \frac{\delta}{h_o}; \quad \frac{E}{G} = 2.6$$

$$\Omega_{1+} = 4\left[1 + \frac{0.3}{L_b^2}\left(\frac{E}{G}\right)\right] L_b^3$$

$$\Omega_{2+} = \frac{6}{\tan^3\sigma} \begin{bmatrix} -2\ln\rho(h_o) - 4H_b\rho(h_o)(1 - L_b\tan\sigma) \\ + H_b(2 + H_b)(1 - L_b\tan\sigma)^2 \rho^2(h_o) \end{bmatrix}$$

$$\Omega_{3+} = -\frac{kE\ln\rho(h_o)}{G\tan\sigma}, \quad \Omega_{1-} = 3L_m(2H_x + 1)L_b,$$

$$\Omega_{2-} = \frac{3L_m}{2\tan\sigma}(2H_x + 1)\left[1 - \rho^2(h_o)\right]$$

Similarly, the net torque about o' is:

$$\begin{bmatrix} T_{X'} \\ T_{Y'} \\ T_{Z'} \end{bmatrix} = \begin{bmatrix} 0 \\ 0 \\ F_{Y'}(h - x - x_0) \end{bmatrix} = \begin{bmatrix} 0 \\ 0 \\ 2\tan\varepsilon_z(h - x - x_0)^2 - 2y(h - x - x_0) \end{bmatrix} \quad (8.18)$$

Ideally, the initial tilting angle during disassembly equals to zero and hence the reaction forces along the Y-axis cancel out. However, if the tilting angle is not zero, an unbalanced reaction force will result and will be felt by the operator through the feedback of the PMSM. As an illustration, a two-phase disassembly process is simulated with values of the simulation parameters given in Table 8.3. The simulated results are discussed as follows:

Phase 1 The plug is initially tilted with $\varepsilon_z = 2°$ and then pulled out along the X' axis. The PMSM is in translational mode. The interaction between the PMSM and the snap-fit assembly as well as the motion command and the force feedback are described in 2nd row of Fig. 8.8a. The results (as shown in 1st row of Fig. 8.8b) indicate that $F_{X'}$ is constant and the non-zero $F_{Y'}$ implies that the plug deviates from the X' axis.

Due to the initially tilted position, the process starts at x_0 with $y = 0$ throughout the simulation. The force feedback is translated into desired

Table 8.3 Simulation parameters

Parameter	Design configuration [2]
Socket (mm)	$\ell_b = 5.4, \ell_m = 9.6, \beta = 58.8°, h_b = 1.8, h_o = 0.6, w = 8, w_s = 4.8$ Young's modulus $E = 1.13$ GPa; Poisson's ratio $\upsilon = 0.41$
Plug (mm)	$w_p = 6.3$, thickness = 1.5 $E = 0.93$ GPa and $\upsilon = 0.35$
μ (friction coefficient)	0.134
Coef. in 3rd term of Eq. (8.12)	$p = 6/5$

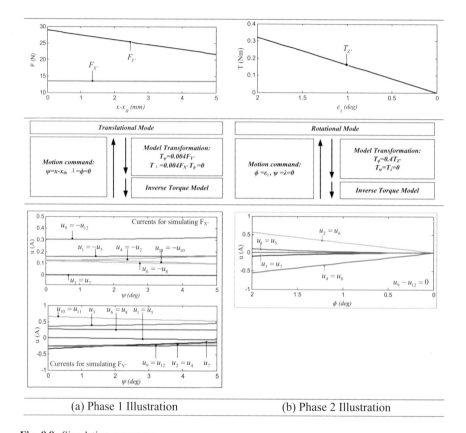

(a) Phase 1 Illustration   (b) Phase 2 Illustration

Fig. 8.8 Simulation responses

torques so that the current inputs can be computed using the inverse torque model where the TCVs are derived using the trained ANN directly from the magnetic fields. For clarification, the current inputs are group into two plots realizing T_ψ (due to $F_{X'}$) and T_λ (due to $F_{Y'}$) as shown in the bottom row of Fig. 8.8a. The resultant torque perceived by a human

8.2 Snap-Fit Simulation

operator through the rotor can be obtained by summing both sets of currents and inputting them into the EMs.

Phase 2 The tilting angle is adjusted by the rotational mode of the PMSM. The interaction leading to the motion command and the force feedback along with the unbalanced reaction torque is described in Fig. 8.8b. As shown in the top row of Fig. 8.8b, the torque decreases as the tilting angle increases and becomes zero when the tilting angle approaches zero. The torque feedback in this process can be realized with the current inputs contributing the desired torque shown in the bottom row of Fig. 8.8b. After the tilting angle is adjusted, the PMSM will be switched to translational mode while the disassembly motion proceeds.

In summary, a two-mode PMSM capable of providing smooth, continuous multi-DOF motion is presented in the context of a haptic application. With the two-mode configuration, the PMSM can offer 6-DOF motion commands for manipulating a target in a virtual computer-aided engineering environment while providing sensible force/torque feedback for human operators. A field-based TCV estimation method using ANN is introduced, which estimates the TCV directly from magnetic fields and permits parallel processing in current input computation. The PMSM haptic device has been numerically demonstrated as an interface between the designer and the virtual design environment for a snap-fit disassembly process. The two-DOF snap-fit haptic applications show that the PMSM offers the human operator an effective means to manipulate targets with multi-DOF force/torque feedback in real time, which greatly improves the performance of such manually operated tasks.

Appendix: PM/EM/Sensor Position Coordinates

The magnetization axes of the PMs or EMs can be characterized by a vector pointing from the origin to the centroid of each PM and EM. The adjacent PMs have alternating magnetization axis. The centroids as well as the sensor positioning are defined in terms of spherical coordinates (as shown in Fig. 8.2a) in rotor frame (for PMs) and stator frame (for EMs and sensors) respectively, which have the form:

Table 8.4 Locations of PMs, EMs and sensors

	Sensor (in XYZ)			EM (in XYZ)			PM (in xyz)	
j	1–16	17–32	33–48	1–8	9–16	17–24	1–12	13–24
ϑ (°)	−26	26	0	−26	26	0	−15	15
φ (°)	22.5 $(j-1)$	22.5 $(j-17)$	22.5 $(j-33)$	45 $(j-1)$	45 $(j-9)+22.5$	45 $(j-17)+22.5$	30 $(j-1)$	30 $(j-13)$

R_{PM} = 67.9 mm, R_{EM} = 56.8 mm, R_{Sensor} = 62.4 mm

$$\vec{r}_j = R \begin{bmatrix} C_{\varphi_j} S_{\vartheta_j} & S_{\varphi_j} S_{\vartheta_j} & C_{\vartheta_j} \end{bmatrix}^T \tag{8.19}$$

The parameters are given in Table 8.4.

References

1. K. Bai, J.J. Ji, K.M. Lee, S.Y. Zhang, A two-mode six-DOF motion system based on a ball-joint-like spherical motor for haptic applications. Comput. Math. Appl. **64**, 978–987 (2012)
2. J.J. Ji, K.M. Lee, Y.S. Zhang, Cantilever snap-fit performance analysis for haptic evaluation. ASME J. Mech. Design **133**(12), 121004(1–8) (2011)